**FOURTH SUPPLEMENT
TO THE
THIRD EDITION**

FOOD CHEMICALS CODEX

Effective March 1, 1994

COMMITTEE ON FOOD CHEMICALS CODEX
Food and Nutrition Board
Institute of Medicine
National Academy of Sciences

NATIONAL ACADEMY PRESS
Washington, D.C. 1993

NATIONAL ACADEMY PRESS **2101 CONSTITUTION AVENUE, NW** **WASHINGTON, DC 20418**

NOTICE   The project that is the subject of this report was approved by the Governing Board of the National Research Council, whose members are drawn from the Councils of the National Academy of Sciences, the National Academy of Engineering, and the Institute of Medicine. The members of the Committee responsible for the report were chosen for their special competences and with regard for appropriate balance.

INSTITUTE OF MEDICINE   The Institute of Medicine was chartered in 1970 by the National Academy of Sciences to enlist distinguished members of the appropriate professions in the examination of policy matters pertaining to the health of the public. In this, the Institute acts under both the Academy's 1863 congressional charter responsibility to be an adviser to the federal government and its own initiative in identifying issues of medical care, research, and education. Dr. Kenneth I. Shine is President of the Institute of Medicine.

FOOD AND NUTRITION BOARD   The Food and Nutrition Board (FNB) was established in 1940 to address issues of national importance that pertain to the safety and adequacy of the nation's food supply; to establish principles and guidelines for adequate nutrition; and to render authoritative judgment on the relationships among food intake, nutrition, and health. The FNB is a multidisciplinary group of scientists with expertise in various aspects of nutrition, nutritional biochemistry, food science and technology, epidemiology, food toxicology, food safety, public health, and food and nutrition policy. These scientists respond to requests from federal agencies about issues concerning food and nutrition, initiate studies that are later assigned to standing or ad hoc FNB committees, and oversee the work of these committees.

Through members of its liaison panels, technical input in aspects of nutrition, food safety, food technology, and food processing is provided.

This study is supported by U.S. Food and Drug Administration Contract No. 223-92-2250.

COMPLIANCE WITH FEDERAL STATUTES   The fact that an article appears in the *Food Chemicals Codex* or its supplements does not exempt it from compliance with requirements of acts of Congress, with regulations and rulings issued by agencies of the United States Government under authority of these acts, or with requirements and regulations of governments in other countries that have adopted the *Food Chemicals Codex*. Revisions of the federal requirements that affect the *Codex* specifications will be included in *Codex* supplements as promptly as practicable.

EFFECTIVE DATE   The specifications in this supplement become effective March 1, 1994.

LIBRARY OF CONGRESS CATALOG NUMBER 93-86131
INTERNATIONAL STANDARD BOOK NUMBER 0-309-04983-0

Copyright 1993 by the National Academy of Sciences. All rights reserved.

No part of this publication may be reproduced by any mechanical, photographic, or electronic process, or in the form of a phonographic recording, nor may it be stored in a retrieval system, transmitted, or otherwise copied for public or private use, without written permission from the publisher, except for official use by the United States Government or by governments in other countries that have adopted the Food Chemicals Codex.

Printed in the United States of America

Contents

ADDITIONS, CHANGES, AND CORRECTIONS, 187

1 GENERAL PROVISIONS APPLYING TO SPECIFICATIONS, TESTS, AND ASSAYS OF THE *FOOD CHEMICALS CODEX,* 188

2 MONOGRAPHS, 189

3 SPECIFICATIONS FOR FLAVOR AROMATIC CHEMICALS AND ISOLATES, 212

4 TEST METHODS FOR FLAVOR AROMATIC CHEMICALS AND ISOLATES, 216

5 GLC ANALYSIS OF FLAVOR AROMATIC CHEMICALS AND ISOLATES, 217

6 GENERAL TESTS AND APPARATUS, 218

7 SOLUTIONS AND INDICATORS, 219

8 GENERAL INFORMATION, 220

9 INFRARED SPECTRA, 221

INDEX, 222

COMMITTEE ON FOOD CHEMICALS CODEX (1991–1993)

Steve L. Taylor, Food Processing Center, University of Nebraska, Lincoln, NE, *Chair*
Samuel M. Tuthill, Mallinckrodt Specialty Chemicals Company, St. Louis, MO, *Vice-Chair*
Herbert Blumenthal, Silver Spring, MD
Joseph T. Brady, Belleville, IL (1982–1992)
Grady William Chism, III, Ohio State University, Department of Food Science and Nutrition, Columbus, OH (1992–present)
Andrew Ebert, The Robert H. Kellen Company, Atlanta, GA
Susan K. Harlander, Department of Food Science and Nutrition, University of Minnesota, St. Paul, MN (1989–1992)
Nancy Higley, Scientific & Regulatory Affairs, Tastemaker, Cincinnati, OH (1992–present)
Joseph H. Hotchkiss, Department of Food Science, Cornell University, Ithaca, NY
Jenny C. Hunter-Cevera, Universal Flavors, Inc., Indianapolis, IN (1992–1993)
B.L. Huston, Chemical Evaluation Division, Health Protection Branch, Ottawa, Ontario, Canada
John C. Kirschman, Emmaus, PA
Francis P. Mahn, Quality Control, Hoffmann-La Roche, Inc., Nutley, NJ
Andrew J. Schmitz, Jr. (deceased), Huntington, NY (1989–1992)
Stephen G. Schulman, College of Pharmacy, University of Florida, Gainesville, FL (1988–1992)
Jan Stofberg, Palmyra, VA (1981–1992)
Connie Marie Weaver, Department of Foods and Nutrition, Purdue University, West Lafayette, IN (1992–present)

Fatima N. Johnson, *Project Director* (1992–present)
Sanford W. Bigelow, *Project Director* (1989–1991)
Sheila A. Moats, *Research Associate*
Geraldine Kennedo, *Project Assistant*

FOOD AND NUTRITION BOARD

M.R.C. Greenwood, *Chair*, Dean of Graduate Studies, University of California, Davis, CA
Edwin L. Bierman, *Vice-Chair*, Division of Metabolism, Endocrinology, and Nutrition, University of Washington, Seattle, WA
Perry L. Adkisson, Department of Entomology, Texas A&M University, College Station, TX
Lindsay Allen, Nutritional Sciences, University of Connecticut, Storrs, CT
Dennis M. Bier, Pediatric Endocrinology and Metabolism, Washington University, St. Louis, MO
Hector F. DeLuca, Department of Biochemistry, University of Wisconsin-Madison, Madison, WI
Michael P. Doyle, Department of Food Science and Technology, University of Georgia, Griffin, GA
Johanna T. Dwyer, Frances Stern Nutrition Center, Boston, MA
John W. Erdman, Jr., Division of Nutritional Sciences, University of Illinois, Urbana, IL
Cutberto Garza, Division of Nutritional Sciences, Cornell University, Ithaca, NY
K. Michael Hambidge, University of Colorado Medical Center, Department of Pediatrics, Denver, CO
Janet C. King, Department of Nutritional Sciences, University of California, Berkeley, CA
John E. Kinsella (deceased), College of Agricultural and Environmental Sciences, University of California, Davis, CA
Laurence N. Kolonel, Cancer Center of Hawaii and University of Hawaii, Honolulu, HI
Sanford Miller, Graduate School of Biomedical Sciences, University of Texas, San Antonio, TX
Alfred Sommer, School of Hygiene and Public Health, Johns Hopkins University, Baltimore, MD
Vernon R. Young, School of Science, Laboratory of Human Nutrition, Massachusetts Institute of Technology, Cambridge, MA

IOM Council Liaison

Arthur H. Rubenstein, Department of Medicine, Division of the Biological Sciences, The University of Chicago, Chicago, IL

Catherine E. Woteki, *Executive Director*

Additions, Changes, and Corrections

Additions, changes, and corrections listed herein constitute revisions in the *Food Chemicals Codex*, Third Edition (FCC III). Page numbers refer to FCC III and its first three supplements unless indicated by a reference to pages in THIS SUPPLEMENT.

1/ *General Provisions Applying to Specifications, Tests, and Assays of the* Food Chemicals Codex

No changes.

2/ Monographs

N-Acetyl-L-Methionine, page 10

Change the *Requirement* entitled *Specific Rotation* to read:

Specific Rotation $[\alpha]_D^{20°}$: Between −18.0° and −22.0°, after drying.

Change the *Test* entitled *Specific Rotation* to read:

Specific Rotation, page 530 Determine in a solution containing 2 g of the previously dried sample in sufficient 2 *N* hydrochloric acid to make 100 mL.

Insert the following new monograph to precede the monograph entitled *Ammonium Alginate*, page 18:

Ammoniated Glycyrrhizin

DESCRIPTION

It is a brown powder. It is precipitated by acid from the water extract of dried and ground rhizomes and roots of *Glycyrrhiza glabra* or related *Glycyrrhiza* (licorice root) and neutralized with dilute ammonia. Suitable diluents may be added.

Functional Use in Foods Flavoring agent; flavor enhancer.

REQUIREMENTS

Identification

It gives positive tests for *Ammonium*, page 515.

Assay Not less than the equivalent of 22.0% and not more than the equivalent of 32.0% of monoammonium glycyrrhizinate, $C_{42}H_{65}NO_{16}$, calculated on the dried basis.
Arsenic Not more than 3 mg/kg.
Ash (Total) Not more than 2.5%.
Heavy Metals (as Pb) Not more than 10 mg/kg.
Loss on Drying Not more than 6.0%.

TESTS

Assay (Based on AOAC method 982.19, 15th Edition, 1990.)
   Apparatus Use a high-pressure liquid chromatograph, operated at room temperature, and containing a 10-µm particle size, 30-cm × 4-mm id, C_{18} reverse-phase column (µBondapak C_{18} or equivalent). Maintain the *Mobile Phase* at a pressure and flow rate (typically 2.0 mL/min) capable of giving the required elution time (see *System Suitability*). Use an ultraviolet detector that monitors absorption at 254 nm (0.2 to 0.1 AUFS range).
   Mobile Phase Add 380 mL of acetonitrile and 10 mL of acetic acid to 610 mL of glass-distilled water filtered through a 0.45-µm filter (Millipore or equivalent). Mix, and de-gas thoroughly.

Standard Solution Weigh accurately about 10 mg of Monoammonium Glycyrrhizinate Standard for analytical use (available from MacAndrews & Forbes Company[1]) in 20 mL of a 1:1 acetonitrile–water solution. Filter the solution through a 0.45-μm Millipore filter or equivalent. Prepare fresh daily. (NOTE: Correct the weight of Monoammonium Glycyrrhizinate Standard taken for the percent loss on drying shown on its label.)

Assay Solution Weigh accurately about 40 mg of the sample, and dissolve in 20 mL of water. Filter the solution through a 0.45-μm Millipore filter or equivalent.

System Suitability Inject duplicate 10-μL portions of the *Standard Solution* into the chromatograph. The retention time of the monoammonium glycyrrhizinate is approximately 6 min. Adjust the operating conditions if necessary. The mean standard deviation for replicate injections is not more than 2.0%.

Procedure Separately inject, in duplicate, 10-μL volumes of the *Standard Solution* and the *Assay Solution* into the chromatograph, and determine the mean peak area for each solution. Calculate the percentage of monoammonium glycyrrhizinate, equivalent to $C_{42}H_{65}NO_{16}$, in the portion of Ammoniated Glycyrrhizin taken by the formula:

$$100(20C_S/W_U)(A_U/A_S),$$

in which C_S is the concentration, in mg/mL, of the *Standard Solution*; A_S and A_U are the peak areas of the *Standard Solution* and the *Assay Solution*, respectively; and W_U is the weight, in mg, of the sample taken.

Arsenic A *Sample Solution* prepared for organic compounds using 1 g of the sample meets the requirements of the *Arsenic Test*, page 464.

Ash (Total) Proceed as directed in the general method, page 466.

Heavy Metals A 2-g sample meets the requirements of the *Heavy Metals Test, Method II*, page 513, using 20 μg of lead ion (Pb) in the control (*Solution A*).

Loss on Drying, page 518 Dry 1 g at 105° for 1 h.

GENERAL INFORMATION

Packaging and Storage Store in a tight container in a cool, dry place.

Annatto Extracts, Second Supplement, page 37

Replace the last sentence of the *Requirement* entitled *Identification A* to read:

Water-soluble Annatto Extract diluted with water exhibits absorbance maxima at 451 to 455 nm and 480 to 484 nm.

[1] Third Street and Jefferson Avenue, Camden, NJ 08104.

Replace the *Requirement* entitled *Identification B* to read:

Carr–Price Reaction Prepare a small chromatography column by filling a glass tube (e.g., 7 × 200 mm), stoppered with glass wool, with alumina (800–200 mesh) slurried in toluene so that the settled alumina fills about 2/3 of the tube. Using a rubber outlet tube and clamp, adjust the flow rate to about 30 drops/min.

Oil-Soluble Annatto Add to the top of the alumina column 3 mL of a solution containing sufficient sample, in toluene, to impart a color equivalent to a 0.1% potassium dichromate solution. Elute with toluene until a pale yellow fraction is washed from the column. Wash the column with three 10-mL volumes of dry acetone, add 5 mL of *Carr–Price Reagent* (page 559), and allow it to run onto the top of the column. The orange-red zone (bixin) at the top of the column immediately becomes blue-green.

Water-Soluble Annatto Transfer 2 mL or 2 g to a 50-mL separatory funnel, and add sufficient 2 N sulfuric acid to make the solution acidic to pH test paper (pH 1 to 2). Dissolve the red precipitate of norbixin by mixing the solution with 50 mL of toluene. Discard the water layer, and wash the toluene phase with water until it no longer gives an acid reaction. Remove any undissolved norbixin by centrifugation or filtration, and dry the solution over anhydrous sodium sulfate. Transfer 3 to 5 mL of the dry solution to the top of an alumina column prepared as described above. Elute the column with toluene, then three 10-mL volumes of dry acetone, followed by 5 mL of *Carr–Price Reagent* added to the top of the column. The orange-red band of norbixin immediately becomes blue-green.

L-Arginine, page 26

Change the *Requirements* entitled *Assay* and *Specific Rotation* to read:

Assay Not less than 98.5% and not more than 101.5% of $C_6H_{14}N_4O_2$, calculated on the dried basis.

Specific Rotation $[\alpha]_D^{20°}$: Between +26.0° and +27.9°, after drying.

Calcium Pantothenate, Calcium Chloride Double Salt, page 57; Third Supplement, page 103

Change the *Test* entitled *Assay* to read:

Assay Proceed as directed for *Assay* under *Calcium Pantothenate*, page 103 of the Third Supplement.

Calcium Pantothenate, Racemic, page 57; Third Supplement, page 103

Change the *Test* entitled *Assay* to read:

Assay Proceed as directed for *Assay* under *Calcium Pantothenate*, page 103 of the Third Supplement.

Insert the following new monograph to precede the monograph entitled *Cognac Oil, Green*, page 89:

Cocoa Butter Substitute

DESCRIPTION

A white, waxy, odorless solid that is predominantly a mixture of triglycerides derived primarily from palm oils. Cocoa Butter Substitute is the common name for the triglyceride consisting mainly of 1-palmitoyl-2-oleoyl-3-stearin. It may be synthesized by the esterification of fully saturated 1,3-diglycerides with the anhydride of food-grade oleic acid in the presence of a catalyst (trifluoromethane sulfonic acid) or by transesterification of partially saturated 1,2,3-triglycerides with ethyl stearate in the presence of a suitable food-grade lipase enzyme preparation approved for such use. The resulting product may be used directly or with cocoa butter in all proportions for the preparation of coatings. In contrast to many edible oils and hard butters, Cocoa Butter Substitute has an abrupt melting range, changing from a rather firm, plastic solid below 32° to a liquid at about 33.8° to 35.5°. Cocoa Butter Substitute is free from any rancid odor and taste.

Functional Use in Foods Coating agent; formulation aid; texturizer.

REQUIREMENTS

Identification

Cocoa Butter Substitute exhibits the following typical composition profile of fatty acids as determined under *Fatty Acid Composition*, page 82 of the Second Supplement.

Fatty Acid:	≤12	12:0	14:0	16:0	16:1
Weight % (Range):	0.0	0.0	0.0	21–24	0.0
Fatty Acid:	18:0	18:1	18:2	≥20	
Weight % (Range):	40–44	31–35	0.5–1.5	0.3–0.7	

Arsenic (as As) Not more than 1 mg/kg.
Color (Lovibond) Not more than 2.5 red.
Free Fatty Acids (as oleic acid) Not more than 1.0%.
Glycerides Not less than 98.0% of total.
   Diglycerides Not more than 7.0%.
   Monoglycerides Not more than 1.0%.
   Triglycerides Not less than 90.0%.
Heavy Metals (as Pb) Not more than 10 mg/kg.
Hexane Not more than 5 mg/kg.
Iodine Value Between 30 and 33.
Lead Not more than 0.1 mg/kg.
Peroxide Value Not more than 10 meq/kg.
Residual Catalyst (as F) Not more than 0.5 mg/kg.
Unsaponifiable Matter Not more than 1.0%.
Water Not more than 0.1%.

TESTS

Arsenic A *Sample Solution* prepared for organic compounds using a 2-g sample, accurately weighed, meets the requirements of the *Arsenic Test*, page 464, using 2.0 mL of *Standard Arsenic Solution* (2 µg As) in the control.

Color Proceed as directed under *Color*, page 82 of the Second Supplement.

Free Fatty Acids Using the diglyceride fraction under *Glycerides*, proceed as directed under *Free Fatty Acids*, page 504, except add 2 mL of phenolphthalein TS, and titrate with the appropriate normality of sodium hydroxide. Use the following equivalence factor (e) in the formula given in the procedure:

Free fatty acids as oleic acid, $e = 28.2$

Glycerides Proceed as directed under *Total Monoglycerides*, page 506, except save all three elution fractions to determine the percentages of *Monoglycerides*, *Diglycerides*, and *Triglycerides*. (NOTE: Use toluene instead of benzene.) The diglyceride fraction also contains free fatty acids, the percentage of which is determined under *Free Fatty Acids*. Calculate the percentage of *Glycerides*, which is the sum of *Monoglycerides*, *Diglycerides*, and *Triglycerides*, by the following formulas:

$$TG = T + D + M,$$

$$T = W_T 100/W_U,$$

$$D = (W_D 100/W_U) - F,$$

$$M = W_M 100/W_U,$$

where TG is the percent of total glycerides; T is the percent of triglycerides; D is the percent of diglycerides; M is the percent of monoglycerides; W_T is the weight, in g, of triglycerides; W_U is the weight, in g, of the sample taken; W_D is the weight, in g, of diglycerides; F is the percent of free fatty acids; and W_M is the weight, in g, of monoglycerides.

Heavy Metals A 2-g sample meets the requirements of the *Heavy Metals Test, Method II*, page 513, using 20 µg of lead ion (Pb) in the control (*Solution A*).

Hexane

Standard Preparation Using a micropipet, transfer and dissolve 34 µL of hexane in 45 g of cold-pressed cottonseed oil (that has not been extracted with hexane). As directed under *Procedure*, analyze aliquots of 0.1, 0.25, 0.5, and 5.0 mg; the aliquots correspond to 2, 5, 10, and 100 mg/kg, respectively, of residual hexane in a 25-mg sample.

Assay Preparation Pack the lower half of 8.5-cm × 9.5-mm (od) borosilicate glass tubing (inlet liner) with glass wool that has been heated at 200° for 16 h to expel volatiles. Transfer a 25-mg sample, accurately weighed, into the glass tubing, and cover it with a small plug of treated glass wool.

Chromatographic System Use a suitable gas chromatograph that is equipped with independent dual flame-ionization detectors and contains a 0.6-m × 6.35-mm (od) stainless-steel U-tube packed with Porapak P or equivalent. After the *Assay Preparation* is inserted into the chromatograph, subject it to the following operating conditions: Maintain the inlet temperature at 110°, maintain the detectors at 200°, and hold the column oven initially at 70° for 2 min followed by a linear temperature gradient at 5°/min for 22 min at a temperature between 70° and 180° and a final hold at 180° for 10 min or until the column is clean. Use helium as the carrier gas at a flow rate of 60 mL/min, hydrogen as the fuel gas at a flow rate of 52 mL/min for each flame, and air as the scavenger gas for both flames at a flow rate of 500 mL/min. To ensure that the relative standard deviation does not exceed 2.0%, chromatograph a sufficient number of replicates of each *Standard Preparation*, and record the areas as directed under *Procedure*. (See *Chromatography*, page 27, First Supplement.)

Standard Curve Chromatograph aliquots of each *Standard Preparation* as directed under *Procedure*. Measure the peak areas for each *Standard Preparation*. Plot a standard curve using the concentration, in mg/kg, of each *Standard Preparation* versus its corresponding peak area, and draw the best straight line.

Procedure Insert the *Assay Preparation* into the inlet liner of the gas chromatograph, immediately sealing the base of the inlet and the lower lip of the glass tubing with a silicone O-ring (Applied Science Laboratories, Inc., or equivalent) previously heated at 200° for 2 h to remove volatile impurities. Immediately close the inlet liner with the septum and septum liner. Allow the carrier gas to flow through the *Assay Preparation*, chromatograph as directed under *Chromatographic System*, and record the chromatograms. Using the peak area of hexane eluting from the *Assay Preparation* at the same time as the *Standard Preparation*, read directly from the *Standard Curve* the concentration, C, of hexane in mg/kg, of the *Assay Preparation*. Calculate the quantity, in mg/kg, of hexane in the sample by the formula:

$$25C/W,$$

in which W is the weight, in mg, of the sample introduced into the gas chromatograph.

Iodine Value Proceed as directed under *Iodine Value*, page 505.

Lead Determine as directed under *Method II* in the *Atomic Absorption Spectrophotometric Graphite Furnace Method* under the *Lead Limit Test*, page 168 of the Third Supplement, using a 5-g sample.

Peroxide Value Proceed as directed under *Peroxide Value*, page 148 of the monograph for *Hydroxylated Lecithin*. However, after the addition of saturated potassium iodide and mixing, instead of allowing the solution to stand for 10 min, mix the solution for 1 min, and begin the titration immediately.

Residual Catalyst Transfer a 30-g sample, accurately weighed, into a 250-mL distillation flask having a side arm and a trap. Connect the flask with a condenser, and fit it with a thermometer and a capillary tube. Both of these should reach nearly to the bottom of the flask so that they extend into the liquid during the distillation. Add 0.2 g of silver sulfate, three boiling beads, and 25 mL of 1:1 sulfuric acid to the flask. Connect a dropping funnel or a steam generator to the capillary tube. Distill until the temperature reaches 135°. Then, through the capillary, add water from the funnel, or introduce steam, as necessary, to maintain the temperature as close to 135° as possible, until 250 mL of distillate has been collected in a beaker. Cool the distillate. Add 3 mL of 30% hydrogen peroxide to remove any sulfites, let stand for 5 min, and evaporate the distillate in a dish containing 15 mL of saturated calcium hydroxide suspension. Ash the residue at 600° for 4 h. Using the ashed residue as the sample, proceed as directed for *Method I* under *Fluoride Limit Test*, page 510, beginning with ". . . and 30 mL of water in a 125-mL distillation flask having a side arm and trap." The total volume of sodium fluoride TS required for the solutions from both *Distillate A* and *Distillate B* should not exceed 0.75 mL.

Unsaponifiable Matter Proceed as directed under *Unsaponifiable Matter*, page 509.

Water Proceed as directed under *Water Determination*, page 552. However, in place of 35 to 40 mL of methanol, use 50 mL of a 1:1 chloroform–methanol mixture to dissolve the sample.

GENERAL INFORMATION

Packaging and Storage Store in well-closed containers.

Cottonseed Oil (Unhydrogenated), Second Supplement, page 42

Change the first sentence of the *Requirement* entitled *Identification* to read:

Unhydrogenated Cottonseed Oil exhibits the following composition profile of fatty acids as determined under *Fatty Acid Composition*, page 82 of the Second Supplement.

Change the *Requirement* entitled *Free Fatty Acids* to read:

Free Fatty Acids (as oleic acid)   Not more than 0.1%.

Delete the rest of the *Requirement* entitled *Free Fatty Acids*.

Change the first sentence of the *Test* entitled *Arsenic* to read:

Arsenic   A *Sample Solution* prepared using 4 g of sample, accurately weighed, meets the requirements of the *Arsenic Test*, page 464, using 2 mL of the *Standard Arsenic Solution* in the control (2 µg As).

Delete the remainder of the *Test* entitled *Arsenic*.

Delete the last line of the *Test* entitled *Free Fatty Acids*.

Change the *Test* entitled *Lead* to read:

Lead   Determine as directed under *Method II* in the *Atomic Absorption Spectrophotometric Graphite Furnace Method* under the *Lead Limit Test*, page 168 of the Third Supplement, using a 3-g sample.

Insert the following new monograph to precede the monograph entitled *Enzyme Preparations*, page 107:

Enzyme-Modified Milkfat

Enzyme-Modified Fat

DESCRIPTION

Light- to medium-tan liquid, paste, or powder with strong fatty acid odor and flavor. Produced by enzyme lipolysis, using suitable food-grade enzymes, of milkfat obtained from the following sources: milk, concentrated milk, dry whole milk, cream, concentrated cream(s), dry cream, butter, butter oil, dried butter, or anhydrous milkfat. Milkfat emulsions are reacted with suitable food-grade enzymes under controlled conditions to increase the flavor components. Optional dairy ingredients such as skim milk, concentrated skim milk, non-fat dry milk, buttermilk, concentrated buttermilk, dried buttermilk, liquid whey, concentrated whey, and dried whey may be used to adjust the concentration of the flavors. Thermoprocessing is then used to destroy the enzyme activity and provide acceptable microbiological quality. Suitable preservatives, emulsifiers, buffers, stabilizers, and antioxidants as well as sodium chloride may be added.

Functional Use in Food   Flavoring agent.

REQUIREMENTS

Identification

Very strong fatty acid odor.

Acid Value   Not less than 98.0% and not more than 102.0% of the labeled value.
Arsenic (as As)   Not more than 1 mg/kg.
Heavy Metals (as Pb)   Not more than 10 mg/kg.
Lead   Not more than 1 mg/kg.
Loss on Drying   Not more than 4.0%.
Microbial Limits:
   Aerobic Plate Count   Not more than 10,000 per g.
   Coliforms   Not more than 10 per g.
   Salmonella   Negative.
   Staphylococci   Negative.
   Yeasts and Molds   Not more than 10 per g.

ADDITIONAL REQUIREMENTS

Labeling   Label to indicate the *Acid Value*.

TESTS

Acid Value   Determine as directed under *Method II* in the general procedure, page 503, using a 5-g sample.
Arsenic   A *Sample Solution* prepared as directed for organic compounds and using a 2-g sample meets the requirements of the *Arsenic Test*, page 464, using 2 mL of the *Standard Arsenic Solution* in the control (2 µg As).
Heavy Metals   Prepare and test a 2-g sample as directed in *Method II* under the *Heavy Metals Test*, page 513, using 20 µg of lead ion (Pb) in the control (*Solution A*).
Lead   Determine as directed under *Method II* in the *Atomic Absorption Spectrophotometric Graphite Furnace Method* under the *Lead Limit Test*, page 168 of the Third Supplement.
Loss on Drying, page 518   Dry at 105° for 48 h.
Microbial Limits:
   Aerobic Plate Count   Proceed as directed in Chapter 3 of the *FDA Bacteriological Analytical Manual*, Seventh Edition, 1992.
   Coliforms   Proceed as directed in Chapter 4 of the

FDA Bacteriological Analytical Manual, Seventh Edition, 1992.

Salmonella Proceed as directed in Chapter 5 of the *FDA Bacteriological Analytical Manual*, Seventh Edition, 1992.

Staphylococci Proceed as directed in Chapter 12 of the *FDA Bacteriological Analytical Manual*, Seventh Edition, 1992.

Yeasts and Molds Proceed as directed in Chapter 18 of the *FDA Bacteriological Analytical Manual*, Seventh Edition, 1992.

GENERAL INFORMATION

Packaging and Storage Store in tight containers in a cool place.

L-Glutamic Acid, page 135

Change the *Requirement* entitled *Specific Rotation* to read:

Specific Rotation $[\alpha]_{546.1\ nm}^{20°}$: Between +37.7° and +38.5°; or $[\alpha]_D^{20°}$: Between +31.5° and +32.5°, after drying.

Change the *Test* entitled *Specific Rotation* to read:

Specific Rotation, page 530 $[\alpha]_{546.1\ nm}^{20°}$: Determine in a solution containing 11.8 g of a previously dried sample in sufficient 1.5 N hydrochloric acid to make 100 mL; or $[\alpha]_D^{20°}$: Determine in a solution containing 10 g of a previously dried sample in sufficient 2 N hydrochloric acid to make 100 mL.

Glyceryl Monostearate, Third Supplement, page 116

Insert the following under *Requirements*:

Identification Heat the sample with 3 parts water to 2° to 5° above its melting point. An irreversible gel forms when the sample is held at this temperature.

Invert Sugar, Second Supplement, page 53

Change the second sentence in the *Description* to read:

Invert Sugar is marketed as Invert Sugar Syrup and also contains dextrose (glucose), fructose, and sucrose in various amounts as represented by the manufacturer.

Change the *Requirement* entitled *Assay* to read:

Assay Not less than 90.0% and not more than 110.0% of the labeled amount of sucrose and of Invert Sugar.

Change the first sentence of the *Test* entitled *Assay, Standard Solution*, to read:

Transfer approximately 9.5 g of USP-grade sucrose, accurately weighed, to a 1000-mL volumetric flask; dissolve in 100 mL of water, add 5 mL of hydrochloric acid, and store 3 d at 20° to 25°.

Change the second formula in the *Test* entitled *Assay, Sucrose*, to read:

$$P_H = 100 C_H / C_S$$

Change the third formula in the *Test* entitled *Assay, Sucrose*, to read:

$$P_S = 0.95(P_H - P_I)$$

Change the *Test* entitled *Total Sugars* to read:

Total Sugars Calculate the *Total Sugars* (T_S) as the sum of the percentages of Invert Sugar (P_I) and sucrose (P_S) determined under *Assay*:

$$T_S = P_I + P_S.$$

Insert the following new monograph to precede the monograph entitled *Lactylated Fatty Acid Esters of Glycerol and Propylene Glycol*, page 160:

Lactose

4-*O*-β-Galactopyranosyl-D-glucose

(α-Lactose)

$C_{12}H_{22}O_{11}$ Anhydrous 342.30

$C_{12}H_{22}O_{11} \cdot H_2O$ Monohydrate 360.31

CAS: [63-42-3]

DESCRIPTION

A white to creamy white crystalline powder, normally obtained from whey, possessing a mildly sweet taste; odorless to slight characteristic odor. It may be anhydrous, contain one molecule of water of hydration, or contain a mixture of both forms if it has been prepared by a spray-drying process. It is soluble in water, very slightly soluble in alcohol, and insoluble in chloroform and in ether.

Functional Use in Foods Nutritive sweetener; formulation aid; processing aid; humectant (anhydrous form); texturizer.

REQUIREMENTS

Identification

Add 5 mL of sodium hydroxide TS to 5 mL of a hot, saturated solution of Lactose, and gently warm the mixture. The liquid becomes yellow and, finally, brownish red. Cool to room temperature, and add a few drops of alkaline cupric tartrate TS. A red precipitate of cuprous oxide is formed.

Assay Not less than 98.0% and not more than 100.5% of $C_{12}H_{22}O_{11}$, calculated on the dried basis.
Arsenic Not more than 0.5 mg/kg.
Heavy Metals (as Pb) Not more than 5 mg/kg.
Lead Not more than 0.5 mg/kg.
Loss on Drying *Monohydrate and spray-dried mixture*: Not less than 4.5% and not more than 5.5%; *anhydrous form*: Not more than 1.0%.
pH Not less than 4.5 and not more than 7.5, in a 10% solution.
Residue on Ignition Not more than 0.3%.
Microbial Limits:
    E. coli Negative.
    Salmonella Negative.

ADDITIONAL REQUIREMENTS

Labeling Label to indicate whether it is anhydrous or the monohydrate or a mixture of both forms if it has been prepared by a spray-drying process.

TESTS

Assay Proceed as directed in the *Lactose Test*, page 85 in the Second Supplement. Use about 2 g of the sample, accurately weighed, and transfer it to a 100-mL volumetric flask. Add 10 mL of fructose internal standard solution, dilute to volume with water, and mix. Perform the analysis within 24 h.
Arsenic A solution of 4 g in 35 mL of water meets the requirements of the *Arsenic Test*, page 464, using 2.0 mL of the Standard Arsenic Solution in the control (2 µg As).
Heavy Metals Prepare and test a 4-g sample as directed in *Method II* under the *Heavy Metals Test*, page 513, using 20 µg of lead ion (Pb) in the control (*Solution A*).
Lead Determine as directed under *Method I* in the *Atomic Absorption Spectrophotometric Graphite Furnace Method* under the *Lead Limit Test*, page 169 of the Third Supplement, using a 2-g sample.
Loss on Drying, page 518 Dry 2 g at 120° for 16 h.
pH Transfer a 10-g sample, accurately weighed, into a clean, dry 100-mL Erlenmeyer flask, and add 90 mL of recently boiled water set at 25°. Shake until the particles are evenly suspended and the mixture is free of lumps. Heat the sample to boiling, and shake frequently to aid dissolution. Let the suspension stand for 10 min, decant the supernate into the hydrogen–ion vessel, and quickly cool to 25°. Immediately determine the pH by the *Potentiometric Method*, page 531, using pH 4.01 and 9.18 buffer solutions to standardize the pH meter.
Residue on Ignition, page 533 Ignite 2 g as directed in *Method I*.
Microbial Limits:
    E. coli Proceed as directed in Chapter 4 of the *FDA Bacteriological Analytical Manual*, Seventh Edition, 1992.
    Salmonella Proceed as directed in Chapter 5 of the *FDA Bacteriological Analytical Manual*, Seventh Edition, 1992.

GENERAL INFORMATION

Packaging and Storage Store in well-closed containers protected from humidity.

Lemon Oil, Coldpressed, page 168

Change the *Test* entitled *Solubility in Alcohol* to read:

Proceed as directed in the general method, page 502. One mL dissolves in 3 mL of 95% alcohol, sometimes with a slight haze.

Insert the following new monograph to precede the monograph entitled *Locust (Carob) Bean Gum*, page 174:

Linoleic Acid

(Z,Z)-9,12-Octadecadienoic Acid

$$CH_3(CH_2)_4CH=CHCH_2CH=CH(CH_2)_7COOH$$

$C_{18}H_{32}O_2$ Mol wt 280.44

CAS: [60-33-3]

DESCRIPTION

An essential fatty acid and the major constituent of many vegetable oils, including cottonseed, soybean, peanut, corn, sunflower seed, safflower, poppy seed, and linseed. It is a colorless to pale-yellow, oily liquid that is easily oxidized by air. Its specific gravity is about 0.901, and its refractive index is about 1.4699. It has a boiling point ranging from 225° to 230° and a melting point around −5°. One mL dissolves in 10 mL of petroleum ether. It is freely soluble in ether; soluble in absolute alcohol and chloroform; and miscible with dimethylformamide, fat solvents, and oils. It is insoluble in water.

Functional Use in Foods   Flavoring adjuvant; dietary supplement.

REQUIREMENTS

Identification

Linoleic Acid exhibits the following composition profile of fatty acids as determined under *Fatty Acid Composition*, page 82 of the Second Supplement.

Fatty Acid:	14:0	16:0	18:0	18:1	18:2	18:3
Weight % (Range):	<1.0	3–5	<1.0	<25.0	>60.0	<9.0

Assay   Not less than 60.0% of fatty acid C18:2 equivalent to $C_{18}H_{32}O_2$, calculated on the anhydrous basis.
Acid Value   Between 196 and 202.
Arsenic (as As)   Not more than 3 mg/kg.
Heavy Metals (as Pb)   Not more than 10 mg/kg.
Iodine Value   Between 145 and 160.
Residue on Ignition   Not more than 0.01%.
Saponification Value   Between 194 and 202.
Unsaponifiable Matter   Not more than 2.0%.
Water   Not more than 0.5%.

TESTS

Assay   Proceed as directed under *Fatty Acid Composition*, page 82 of the Second Supplement.
Acid Value   Determine as directed under *Method I* in the *General Procedure*, page 503.
Arsenic   A *Sample Solution* prepared as directed for organic compounds meets the requirements of the *Arsenic Test*, page 464.
Heavy Metals   Prepare and test a 2-g sample as directed in *Method II* under the *Heavy Metals Test*, page 513, using 20 μg of lead ion (Pb) in the control (*Solution A*).
Iodine Value   Determine by the *Wijs Method*, page 505.
Residue on Ignition   Ignite 10 g as directed in the general method, page 533.
Saponification Value   Determine as directed in the general method, page 509, using a 3-g sample, accurately weighed.
Unsaponifiable Matter   Proceed as directed under the general method, page 509, using a 10-g sample.
Water   Determine by the *Karl Fischer Titrimetric Method*, page 552.

GENERAL INFORMATION

Packaging and Storage   Store in tight containers.

Locust (Carob) Bean Gum, page 174

Change the *Requirements* entitled *Acid-Insoluble Matter*, *Loss on Drying*, *Galactomannans*, and *Protein* to read:

Acid-Insoluble Matter   Not more than 4.0%.
Loss on Drying   Not more than 14.0%.
Galactomannans   Not less than 75.0%.
Protein   Not more than 7.0%.

L-Lysine Monohydrochloride, page 176

Change the *Requirements* entitled *Loss on Drying* and *Specific Rotation* to read:

Loss on Drying   Not more than 1.0%.
Specific Rotation $[\alpha]_D^{20°}$: Between +20.5° and +21.5°, after drying.

Change the *Test* entitled *Specific Rotation* to read:

Specific Rotation, page 530   Determine in a solution containing 8 g of a previously dried sample in sufficient 6 *N* hydrochloric acid to make 100 mL.

Insert the following new monograph to precede the monograph entitled *Mandarin Oil, Coldpressed*, page 185:

Malt Syrup

Malt Extract

CAS: [8002-48-0]

DESCRIPTION

Malt is the product of barley (*Hordeum vulgare* L.) germinated under controlled conditions. Malt Syrup and Malt

Extract are interchangeable terms for a concentrate of the water extract of germinated barley grain with or without added food-grade preservatives. Malt Syrup is usually a yellow to brown, sweet, and viscous liquid containing varying amounts of amylolytic enzymes and plant constituents.

Functional Use in Foods Color; enzyme; flavoring agent; humectant; nutritive sweetener; stabilizer; thickener; and texturizer.

REQUIREMENTS

Identification

To 5 mL of hot alkaline cupric tartrate TS add a few drops of a 1 in 10 solution of the sample. A red precipitate of cuprous oxide is formed.

Assay Not less than 40.0% and not more than 65.0% of reducing sugar content expressed as maltose.
Arsenic Not more than 3 mg/kg.
Heavy Metals Not more than 5 mg/kg.
Lead Not more than 0.5 mg/kg.
N-Nitrosodimethylamine Not more than 0.005 mg/kg.
pH Between 4.5 and 5.5.
Protein Not more than 7.0%.
Sulfur Dioxide Not more than 10 mg/kg.
Total Solids Between 77.0% and 83.0%.

TESTS

Assay

Sample Solution Transfer about 5 g, accurately weighed, of the sample to a 500-mL volumetric flask, dilute with water, and mix. (It is not necessary to remove any protein before analysis.) Pipet 10.0 mL each of *Fehling Solutions A* and *B* into a 250 mL flask. Add 20.0 mL (choose the size of the aliquot so that the sample titration will be about half that of the blank titration) of the *Sample Solution*, add water to make a total volume of 50 mL, and mix the contents of the flask by gentle swirling. Add two small glass beads, and close the mouth of the flask with a small funnel or glass bulb. Heat the solution, preferably on a hot plate, at such a rate that the solution is brought to boiling within 3 min, and then continue boiling for exactly 2 min (total heating time, 5 min). Cool quickly to room temperature in an ice bath or with cold running water, and then rinse down the funnel (or bulb) and the walls of the flask with a few mL of water. Add 10 mL each of 30% potassium iodide solution and 28% sulfuric acid, and titrate rapidly with 0.1 N sodium thiosulfate until the iodine color almost disappears. Add 1 mL of starch TS, and titrate dropwise, with continuous agitation, to the disappearance of the blue color. Record the volume, in mL, of 0.1 N sodium thiosulfate required for the *Sample Solution* as S. Conduct two reagent blank determinations, substituting 20 mL (or the same volume as the aliquot of the *Sample Solution* taken) of water for the sample, and record the average volume, in mL, of the blanks as B. Obtain the *Titer Difference*, expressed as mL

Conversion of Titer Differences to Reducing Sugars Content

Titer Diff. (mL)	.0	.1	.2	.3	.4	.5	.6	.7	.8	.9
					Reducing Sugars (as Maltose) (mg)					
5.0	27.0	27.6	28.1	28.7	29.2	29.8	30.3	30.9	31.4	32.0
6.0	32.5	33.1	33.6	34.2	34.7	35.3	35.8	36.3	36.9	37.5
7.0	38.0	38.6	39.1	39.7	40.2	40.8	41.3	41.9	42.4	43.0
8.0	43.5	44.1	44.6	45.2	45.7	46.3	46.8	47.4	47.9	48.5
9.0	49.0	49.6	50.2	50.8	51.4	52.0	52.6	53.2	53.8	54.4
10.0	55.0	55.6	56.1	56.7	57.2	57.8	58.3	58.9	59.4	60.0
11.0	60.5	61.1	61.6	62.2	62.7	63.3	63.8	64.4	64.9	65.5
12.0	66.0	66.6	67.2	67.8	68.4	69.0	69.6	70.2	70.8	71.4
13.0	72.0	72.6	73.2	73.8	74.4	75.0	75.6	76.2	76.8	77.4
14.0	78.0	78.6	79.1	79.7	80.2	80.8	81.3	81.9	82.4	83.0
15.0	83.5	84.1	84.6	85.2	85.7	86.3	86.8	87.4	87.9	88.5
16.0	89.0	89.6	90.2	90.8	91.4	92.0	92.6	93.2	93.8	94.4
17.0	95.0	95.6	96.2	96.8	97.4	98.0	98.6	99.2	99.8	100.4
18.0	101.0	101.6	102.2	102.8	103.4	104.0	104.6	105.2	105.8	106.4
19.0	107.0	107.6	108.1	108.7	109.2	109.8	110.3	110.9	111.4	112.0
20.0	112.5	113.1	113.7	114.3	114.9	115.5	116.1	116.7	117.3	117.9

of 0.1 N sodium thiosulfate, for the sample by subtracting S from B and recording the value thus obtained as T_S.

Calculation By reference to the preceding table, determine the weight, in mg, of reducing sugars (as maltose) equivalent to the volume T_S, and record the value thus obtained as W_S. Calculate the total reducing sugars (as maltose), in mg, in the sample taken by the formula $25W_S$. (If the aliquot of the *Sample Solution* taken for assay differs from 20.0 mL, adjust the factor accordingly.)

Arsenic A *Sample Solution* prepared as directed for organic compounds meets the requirements of the *Arsenic Test*, page 464.

Heavy Metals Prepare and test a 4-g sample as directed in *Method II* under the *Heavy Metals Test*, page 513, using 20 µg of lead ion (Pb) in the control (*Solution A*).

Lead Determine as directed under *Method I* in the *Atomic Absorption Spectrophotometric Graphite Furnace Method* under the *Lead Limit Test*, page 168 of the Third Supplement, using a 2-g sample.

N-Nitrosodimethylamine (Based on AOAC method 982.12, 15th Edition, 1990.)

Caution: N-Nitrosamines are potent carcinogens: Take adequate precaution to avoid exposure. Carry out all steps in a well-ventilated fume hood, and wear protective gloves while handling nitrosamine standards. Because these compounds are highly photolabile, carry out all procedures under subdued light. Do not pipet solutions by mouth, and do not use the same pipet for other reagents. Destroy all nitrosamine solutions by boiling with hydrochloric acid, potassium iodide, and sulfamic acid before disposal.

NOTE: Thoroughly clean all glassware before use. After normal cleaning and washing, wash with chromic acid. If contamination still exists, rinse all glassware with dichloromethane before use. Let the charred residue in the distillation flask soak with diluted alkali, and then wash in a normal manner.

Apparatus Set up the distillation apparatus consisting of a 1000-mL distillation flask, a heating mantle, an adapter, and a 200-mm Graham condenser (Kontes or equivalent) so that the connecting adapter slopes downward toward the vertical Graham condenser. Loosely wrap glass wool around the distillation flask and connecting adapter. Set up a 100-mL graduate under the condenser to collect the distillate. The cooling water for the condenser should be ≤20°. Assemble a 250-mL Kuderna–Danish evaporative concentrator that has a 24/40 column connection and a 19/22 lower joint (Kontes or equivalent) with a 4-mL Kuderna–Danish concentrator tube (Kontes or equivalent) that has a 19/22 joint and 0.1-mL subdivisions from 0 to 2.0 mL at the bottom.

Use a gas chromatograph fitted with a 6-ft × 1/8-in. (od) stainless steel column packed with 20% Carbowax 20M (or equivalent) and 2% sodium hydroxide on 80- to 100-mesh acid-washed Chromosorb P (or equivalent). Set the temperature of the injector port and column to 220° and 170°, respectively, and use argon as the carrier gas at a flow rate of 25 to 30 mL/min. Use a Thermal Energy Analyzer detector (Thermo Electron Corporation, Waltham, MA, or equivalent). Operate according to the instrument manual and with a –110° to –130° slush bath. Adjust instrumental parameters, such as vacuum chamber pressure, oxygen flow, calibration knob, etc., to obtain the proper sensitivity.

NDMA Standard Solution Using an accurately weighed quantity of N-nitrosodimethylamine (NDMA) (Sigma Chemical Company, St. Louis, MO), prepare a stock solution in dichloromethane having a concentration of 1 mg/mL. By serial dilution with dichloromethane, prepare a series of solutions containing 500, 200, 100, 40, 20, 10, and 5 ng/mL. Store these solutions at –20°, and warm to room temperature before use. After 30 days, dispose of the *Standard Solutions*.

NDPA Standard Solution As described above, prepare a solution containing 250 ng of N-nitrosodi-n-propylamine (NDPA) per mL of anhydrous ethanol.

Sample Preparation Transfer 50 g of the sample, accurately weighed, into a 1000-mL distillation flask, and add 1.0 mL of 10% sulfamic acid, 1.0 mL of *NDPA Standard Solution*, 1.0 mL of 1 N hydrochloric acid, and 15 mL of water. Mix the contents by gentle swirling, and let the flask stand in the dark for 10 min. Add 10.0 mL of 3 N potassium hydroxide and two small boiling chips, mix, and connect the flask to the distillation apparatus.

Distillation During the initial 10 min of distillation, adjust the heating mantle so the mixture boils smoothly without too much frothing or bumping. Watch constantly for excessive foaming, and if necessary, turn off the heat for 1 to 2 min. After 10 min, increase the temperature, and continue distillation (watch for foaming) until approximately 55 mL of distillate is collected. Do not boil the distilling flask to complete dryness; this may give erroneous results. Total distillation time should be ≤1 h. If any portion of the sample foams over during distillation, discontinue the distillation, and start over with a fresh sample of Malt Syrup.

Add 2.0 mL of 10 N potassium hydroxide to the distillate in the graduate, and transfer to a 250-mL separatory funnel. Use the same cylinder for all subsequent measuring of dichloromethane. Rinse the condenser with 50 mL of dichloromethane, and collect the rinsing directly into the separatory funnel containing the distillate and potassium hydroxide. Extract the distillate with dichloromethane by shaking vigorously for 2 min. Drain off the dichloromethane layer (lower) into a second separatory funnel. Extract the aqueous layer with two additional 50-mL portions of dichloromethane, and combine all dichloromethane extracts in the second separatory funnel. Discard the aqueous layer.

Place 40 g of anhydrous sodium sulfate in a coarse, sintered-glass buchner funnel, wash with about 20 mL of dichloromethane, and discard the washing. Dry the combined dichloromethane extract by passing it through a sodium sulfate bed on the buchner funnel, and collect the extract directly in the Kuderna–Danish concentrator. Wash the sodium sulfate bed with an additional 20 mL of dichloromethane, and collect the washing in the Kuderna–Danish concentrator.

Add a 1- to 2-mm boiling chip to the contents of the Kuderna–Danish concentrator, attach a three-section Snyder column with three chambers and a 24/40 joint (Kontes or equivalent), and concentrate the extract by heating the flask in a 50° to 60° water bath. Initially maintain the outside water level close to the level of dichloromethane inside the flask, and continue heating until the concentrated extract is about 4 mL (about 40 min). (If excessive boiling occurs during concentration, control it either by raising the flask slightly out of the water bath or by decreasing the bath temperature.) Finally, raise the flask above the water, and let condensed dichloromethane in the Snyder column drain into the flask. Add about 1 mL of dichloromethane to the top of the Snyder column, and let it drain into the flask. Disconnect the concentrator tube from the flask.

Add another boiling chip to the contents and attach a micro Snyder column with three chambers and a 19/22 joint (Kontes or equivalent) to the concentrator tube. Concentrate the extract to about 0.8 mL by heating the concentrator tube in a 50° to 60° water bath. Lift out or immerse the tube in water to control the boiling rate, but do not lift the tube completely out of the water bath; this will stop the action of the boiling chip. Avoid overheating and excessive accumulation of dichloromethane in the column chambers. Stop concentration when the dichloromethane level reaches 0.8 mL; do not concentrate to less than 0.8 mL. Carry out this final concentrating step slowly, taking at least 30 min. Raise the tube, and with the bottom still touching the water, let the liquid drain, and note the volume to see if it is around 0.8 mL. If it is greater than 0.8 mL, continue the concentration as above. Finally, rinse the micro Snyder column with a few drops of dichloromethane, let the rinsing drain to the tube, disconnect the column, and dilute the extract to 1.0 or 1.1 mL, but not greater than 1.1 mL. (Do not use a nitrogen stream for concentrating the extract at any stage.)

Stopper the tube, mix in a vortex mixer, and store at 4° in the dark until analysis. Let the extract warm to room temperature, and note its volume before analyzing it.

To ensure the absence of contamination, carry the reagent blank taken through all of the steps mentioned above, except use 50 mL of 4% alcohol in water instead of 50 g of the sample.

Procedure Set attenuation (usually 4) of the thermal energy analyzer detector so that injection of 30 pg of NDMA gives a definite peak with acceptable background. Using this attenuation, analyze 5- to 6-µL aliquots, in duplicate, of *NDMA Standard Solutions* of 5, 10, 20, and 40 ng/mL. (Note the volume injected.)

Next, choose a higher attenuation setting that gives an on scale peak for 6 µL of *NDMA Standard Solution* at 500 ng/mL. Using this setting, analyze 6-µL aliquots, in duplicate, of *NDMA Standard Solutions* of 500, 200, 100, and 40 ng/mL.

Accurately measure the peak heights (±0.1 cm), and determine the average peak heights of two injections at each concentration. If exactly 6 µL are not injected, make appropriate corrections, and convert all peak heights equivalent to 5.0-µL injections. Draw two standard curves, one for each attenuation setting, of peak heights versus pg injection.

As above, inject a 6-µL aliquot of sample extract, in duplicate, using the lowest attenuation setting sensitive to 30 pg of NDMA. Measure and determine the average peak height. Compare the sample response with the standard curve that produces the closest peak height at the same attenuation. Choose the *NDMA Standard Solution* that gives the closest peak height, inject 6-µL aliquots, in duplicate, and determine the average peak height.

For samples giving off-scale peaks at an attenuation of 32, dilute the extracts with dichloromethane to 5.0 mL in a volumetric flask, and reanalyze. For accurate results, analyze the sample extract and corresponding standard under the same attenuation setting, and all within 60 min.

If the extract gives a negative result for NDMA, or if the peak is too small to measure, inject 10-µL aliquots, in duplicate, using a 25-µL syringe. Similarly, inject duplicate 10-µL aliquots of the 5 ng/mL *NDMA Standard Solution* for quantitation. To achieve a 0.1-µg/kg detection limit, 10-µL aliquots of *Sample Preparation* must be analyzed under an attenuation setting that gives a detectable peak corresponding to 30 pg of NDMA.

NOTE: If using a 25-µL syringe, which usually has a thick needle, watch for septum damage, and check for leaks. To be on the safe side, use a new septum daily.

Calculate the concentration of NDMA in the sample using the following formula:

Uncorrected NDMA, in µg/kg, in the sample =
$(H_1PV_2)/(H_2GV_1)$,

where H_1 is the average NDMA peak height, in cm, of the sample; H_2 is the average peak height, in cm, of the corresponding *NDMA Standard Solution*; P is the weight, in pg, of NDMA producing the H_2 peak height; V_1 is the volume, in µL, of *Sample Preparation* injected; V_2 is the final volume, in mL, of *Sample Preparation*; and *G* is the weight, in g, of the sample taken for analysis.

Correction for Percent Recovery of NDPA Accurately measure the peak height of the NDPA peak on each sample chromatogram and calculate the average peak height of two injections. Make appropriate corrections if the final volume of sample is not exactly 1.0 mL or the injection volume is not exactly 6.0 µL. Then inject, in duplicate, within 60 min, 6-µL of the *NDPA Standard Solution* under the same attenuation setting. Calculate the average peak height, and correct the value if exactly 6.0 µL is not injected. Calculate the percent recovery of NDPA for each sample. If recovery of NDPA is less than 80%, repeat the analysis from the beginning. Finally, correct the results as follows:

Corrected NDMA, in mg/kg, in the sample = (uncorrected µg/kg/% recovery of NDPA) × 0.1.

pH Determine by the potentiometric method, page 531, using a 1 in 10 aqueous solution.

Protein Using a 0.25-g sample, determine the percent of nitrogen as directed under *Nitrogen Determination (Kjeldahl Method)*, page 521. The percent of nitrogen, multiplied by 6.25, gives the percent of protein in the sample.

Sulfur Dioxide Determine as directed in the general method, page 170 of the Third Supplement, using a 100-g sample.

Total Solids Determine the water content of an accurately weighed portion of the sample by the *Karl Fischer Titrimetric Method*, page 552. Calculate the percent of *Total Solids* by the formula

$$100[(W_U - W_W)/W_U],$$

in which W_U is the weight, in mg, of the sample, and W_W is the weight, in mg, of water determined.

GENERAL INFORMATION

Packaging and Storage Store in tight containers.

Insert the following new monograph to precede the monograph entitled *Mono- and Diglycerides*, page 201.

Monoammonium Glycyrrhizinate

Ammonium Glycyrrhizinate, Pentahydrate; Ammonium Glycyrrhizinate

$$C_{42}H_{61}O_{16}NH_4 \cdot 5H_2O$$

$C_{42}H_{65}NO_{16} \cdot 5H_2O$ Mol wt (anhydrous) 839.91

INS: 958 CAS: [1407-03-0]

DESCRIPTION

It is a white powder with an intensely sweet taste and is obtained by extraction from ammoniated glycyrrhizin. It is soluble in ammonia water and is insoluble in glacial acetic acid.

Functional Use in Foods Flavoring agent.

REQUIREMENTS

Identification

It gives positive tests for *Ammonium*, page 515.

Assay Not less than 85.0% and not more than 102.0% of $C_{42}H_{65}NO_{16}$, calculated on the dried basis.
Arsenic Not more than 3 mg/kg.
Ash (Total) Not more than 0.5%.
Heavy Metals (as Pb) Not more than 10 mg/kg.
Loss on Drying Not more than 6.0%.
Specific Rotation $[\alpha]_D^{20°}$: Between +45° and +53° on the as-is basis.

TESTS

Assay (Based on AOAC method 982.19, 15th edition, 1990.)

Apparatus Fit a high-pressure liquid chromatograph, operated at room temperature, with a 10-µm particle size, 30-cm × 4-mm id, C_{18} reverse-phase column (µBondapak C_{18} or equivalent). Maintain the *Mobile Phase* at a pressure and flow rate (typically 2.0 mL/min) capable of giving the required elution time (see *System Suitability*). Use an ultraviolet detector that monitors absorption at 254 nm (0.2 to 0.1 AUFS range).

Mobile Phase Add 380 mL of acetonitrile and 10 mL of acetic acid to 610 mL of glass-distilled water filtered through a 0.45-µm filter (Millipore or equivalent). Mix, and de-gas thoroughly.

Standard Solution Weigh accurately about 10 mg of Monoammonium Glycyrrhizinate Standard for analytical use (available from MacAndrews & Forbes Company[2]) in 20 mL of a 1:1 acetonitrile–water solution. Filter the solution through a 0.45-µm Millipore filter or equivalent. Prepare fresh daily. (NOTE: Correct the weight of Monoammonium Glycyrrhizinate Standard taken for the percent loss on drying shown on its label.)

Assay Solution Weigh accurately about 10 mg of the sample, and dissolve in 20 mL of a 1:1 acetonitrile–water solution. Filter the solution through a 0.45-µm Millipore filter or equivalent.

---

[2] Third Street and Jefferson Avenue, Camden, NJ 08104.

System Suitability Inject duplicate 10-μL portions of the *Standard Solution* into the chromatograph. The retention time of the Monoammonium Glycyrrhizinate is approximately 6 min. Adjust the operating conditions if necessary. The mean standard deviation for replicate injections is not more than 2.0%.

Procedure Separately inject, in duplicate, 10-μL volumes of the *Standard Solution* and the *Assay Solution* into the chromatograph, and determine the mean peak area for each solution. Calculate the percentage of Monoammonium Glycyrrhizinate, equivalent to $C_{42}H_{65}NO_{16}$, in the portion of Monoammonium Glycyrrhizinate taken by the formula:

$$100(20C_S/W_U)(A_U/A_S),$$

where C_S is the concentration, in mg/mL, of the *Standard Solution*; A_S and A_U are the peak areas of the *Standard Solution* and the *Assay Solution*, respectively; and W_U is the weight, in mg, of the sample taken.

Arsenic A *Sample Solution* prepared for organic compounds using 1 g of the sample meets the requirements of the *Arsenic Test*, page 464.

Ash (Total) Proceed as directed in the general method, page 466.

Heavy Metals A 2-g sample meets the requirements of the *Heavy Metals Test, Method II*, page 513, using 20 μg of lead ion (Pb) in the control (*Solution A*).

Loss on Drying, page 518 Dry 1 g at 78° for 4 h under 1-mm Hg vacuum.

Specific Rotation, page 530 Determine in a solution containing a 1.5-g undried sample in sufficient 40% ethanol to make 100 mL.

GENERAL INFORMATION

Packaging and Storage Store in a tight container in a cool, dry place.

Monosodium L-Glutamate, page 203; Third Supplement, page 126

In the Third Supplement, change the *Requirement* entitled *Specific Rotation* to read:

Specific Rotation $[\alpha]_D^{20°}$: Not less than +24.8° and not more than +25.3°; or $[\alpha]_{546.1\ nm}^{20°}$: Not less than +29.7° and not more than +30.2°.

Insert the following new monographs to precede the monograph entitled *Myristic Acid*, page 204:

Morpholine

Tetrahydro-2H-1,4-oxazine; Diethylene Oximide; Diethylene Imidoxide

C_4H_9NO Mol wt 87.12

CAS: [110-91-8]

DESCRIPTION

A clear, colorless, mobile, hygroscopic liquid with a characteristic amine odor. It is miscible with water with the evolution of some heat. It is also miscible with acetone, benzene, ether, castor oil, methanol, and alcohol as well as in many oils such as linseed and pine. (*Caution*: Dermal irritant.)

Functional Use in Foods Boiler water additive.

REQUIREMENTS

Identification

The infrared absorption spectrum of a neat dispersion of the sample between two sodium chloride plates exhibits maxima at the same wavelengths as the typical spectrum, as shown in Section 9, *Infrared Spectra, Series C*, page 221 of THIS SUPPLEMENT.

Assay Not less than 99.0%.
Distillation Range Between 126.0° and 130.0°.
Heavy Metals (as Pb) Not more than 1 mg/kg.
Refractive Index Between 1.454 and 1.455 at 20°.
Specific Gravity Between 0.994 and 0.997 at 25°.

TESTS

Assay

Mixed Indicator Solution Prepare separate 0.1% solutions of bromocresol green in methanol and the sodium salt of methyl red in water. Mix 5 parts by volume of the bromocresol green solution with one part of the methyl red solution.

Transfer 50 mL of water to a 250-mL flask. Add 0.4 mL of *Mixed Indicator Solution*, and neutralize by the drop-wise addition of 0.1 N hydrochloric acid just to the disappearance of the green color. Weigh accurately 1.4 to 1.6 g of sample into the flask, and swirl to effect complete solution. Titrate with 0.5 N hydrochloric acid to the disappearance of the green color. Each mL of 0.5 N hydrochloric acid is equilavent to 43.56 mg of C_4H_9NO.

Distillation Range Proceed as directed in the general method, page 478.

Heavy Metals Evaporate about 20.0 g of the sample to dryness on a steam bath in a glass evaporating dish. Cool, add 2 mL of hydrochloric acid, and slowly evaporate to dryness again on the steam bath. Moisten the residue with 1 drop of hydrochloric acid, add 10 mL of hot water, and digest for 2 min. Cool, and dilute to 25 mL with water. This solution meets the requirements of the *Heavy Metals Test*, page 512, using 20 µg of lead ion (Pb) in the control.

Refractive Index Determine as directed in the general method, page 533.

Specific Gravity Determine at 25° by any reliable method (page 3).

GENERAL INFORMATION

Packaging and Storage Store in tight containers.

Mustard Oil

DESCRIPTION

The volatile oil obtained by the steam and water distillation of the comminuted press-cakes of the seeds from *Brassica nigra* (Linnaeus) W.D.J. Koch or *Brassica juncea* (Linnaeus) Czernjajev. The essential oil forms upon maceration of the comminuted seeds in warm water that releases sinigrin, a β-glucopyranoside, which is subsequently enzymatically hydrolyzed to allyl isothiocyanate. The oil is a clear, pale yellow liquid with a sharp, pungent taste. (*Caution*: Mustard Oil is a lachrymator.)

Functional Use in Foods Flavoring agent.

REQUIREMENTS

Identification

The infrared absorption spectrum of the sample exhibits relative maxima (that may vary in intensity) at the same wavelengths (or frequencies) as those shown in the respective spectrum for allyl isothiocyanate under *Infrared Spectra of Essential Oils*, page 678, using the same test conditions as specified therein.

Assay Not less than 93.0%, as C_3H_5NCS (allyl isothiocyanate).

Refractive Index Between 1.524 and 1.534 at 20°.

Specific Gravity Between 1.008 and 1.019.

TESTS

Assay Determine as directed under *Assay by Gas–Liquid Chromatography M–8a*, page 424.

Refractive Index Determine as directed under *Refractive Index*, page 533, using 20° as the determination temperature.

Specific Gravity Determine by any reliable method, page 3.

GENERAL INFORMATION

Packaging and Storage Store in tight containers in a cool, dry place protected from light.

Insert the following new monographs to precede the monograph entitled *Nutmeg Oil*, page 206:

Nickel

Nickel Catalysts

Ni   Atomic Wt 58.69

                                                CAS: [7440-02-0]

DESCRIPTION

Nickel metal is a lustrous, white, hard, ferromagnetic, metallic solid with no characteristic odor. Nickel is commonly used as a catalyst for hydrogenation reactions for food chemicals. Depending on the use, Nickel catalysts fall into two general categories: *Sponge Nickel Catalyst* and *Supported Nickel Catalyst*.

Sponge Nickel Catalyst is typically used in the manufacture of amines and polyols. It is prepared by chemically treating a Nickel–aluminum amalgam with sodium hydroxide to remove the majority of aluminum, thus leaving a highly porous (skeletal) Nickel solid. The resulting *Sponge Nickel Catalyst* is extremely pyrophoric in air and must be stored under an inert liquid.

Supported Nickel Catalyst is typically used in the manufacture of edible oils. It is prepared from a Nickel salt deposited onto an inert carrier consisting of various types of acceptable silicas, aluminas, or combinations thereof. The Nickel salt-carrier complex is not catalytically active and is converted to *Supported Nickel Catalyst* in a stream of hydrogen at elevated temperatures. After activation, *Supported Nickel Catalyst* is also pyrophoric and must be protected from air typically by suspending it in a food-grade stearine. It usually is supplied in the form of droplets or flakes.

Functional Use in Foods Catalyst for hydrogenation reactions.

REQUIREMENTS

Identification

Sponge Nickel Catalyst Dissolve approximately 100 mg in about 2 mL of hydrochloric acid, and dilute to about 20 mL with water. Place 5 mL of this solution in a test tube, add a few drops of bromine water, and make it slightly alkaline with ammonium hydroxide. Add 2 to 3 mL of a 1% solution of dimethylglyoxime in alcohol. An intense red color or precipitate develops.

Supported Nickel Catalyst Ash as described under *Assay*. Transfer a 5-mL aliquot of the ashed sample solution to a test tube, and complete the identification as described above.

Assay

Sponge Nickel Catalyst: Not less than 83.0% Ni on the as-is basis. (*Caution*: Handle with extreme care—Sponge Nickel is pyrophoric when dried.)

Supported Nickel Catalyst: Not less than 20.0% Ni and not more than 27.0% Ni on the as-is basis.

TESTS

Assay

Sponge Nickel Catalyst Weigh accurately about 500 mg of wet sample, place it into a 400-mL beaker, and dissolve it in 20 mL of 1:1 hydrochloric acid, heating if necessary. Dilute to 250 mL with water. Add 2 g of tartaric acid, heat to about 80°, and add 30 mL of a 1% solution of dimethylglyoxime in alcohol. Add ammonium hydroxide until the solution is slightly ammoniacal, and place the mixture on a steam bath for 20 min. Filter the precipitate into a tared, fritted glass, medium-porosity filter crucible, and wash with hot water until the filtrate is free of chloride. Dry the precipitate at 120° for 2 h, and then to constant weight, and weigh. Calculate the percent Nickel by the following formula:

$$(W_P 20.32)/W_S,$$

in which W_P is the weight, in g, of the precipitate; 20.32 is the percent of Nickel in the precipitate; and W_S is the weight, in g, of the sample taken.

Supported Nickel Catalyst Fill a 100-mL porcelain crucible half-full of ashless filter paper pulp. Accurately weigh 2 g of the finished catalyst, in droplet or flake form, and place it on top of the paper pulp. Transfer the crucible to a muffle furnace set at room temperature, and slowly raise the temperature to 650° so that the stearine melts into the paper, and the organic mass burns and chars slowly. Continue heating at 650° for 2 h or until the carbon is burned off. Then cool, add 20 mL of hydrochloric acid, quantitatively transfer the solution or suspension to a 400-mL beaker, and carefully evaporate to dryness on a steam bath. Cool, add 20 mL of hydrochloric acid, warm to aid dissolution (catalysts containing silica will not dissolve completely), transfer to a 500-mL volumetric flask, dilute to volume with water, and mix. Allow any solids to settle, pipet a clear, 50-mL aliquot into a 400-mL beaker, and dilute to 250 mL with water. (If there is suspended matter in the volumetric flask, filter a portion through a dry, medium-speed filter paper into a dry receiver, and pipet from the receiver.) Continue as directed above, beginning with "Add 2 g of tartaric acid. . . ."

Calculate the percent Nickel by the following formula:

$$10(W_P 20.32)/W_S,$$

in which W_P is the weight, in g, of the precipitate; 20.32 is the percent of Nickel in the precipitate; and W_S is the weight, in g, of the sample taken.

GENERAL INFORMATION

Packaging and Storage

Sponge Nickel Catalyst: Store under liquids such as water, alcohol, or methylcyclohexane, in a cool, dry place.

Supported Nickel Catalyst: Store in tight containers in a cool, dry place.

Nisin Preparation

contains 34 amino acids and has an approximate empirical formula of
$C_{143}H_{228}O_{37}N_{42}S_7$ Approximate Mol wt 3348

INS: 234   CAS: [1414-45-5]

DESCRIPTION

Nisin is a mixture of closely related polypeptides produced by strains of the *Lactococcus lactis* subsp. *lactis* Lancefield Group N in a sterilized milk-culture medium. Nisin in the fermentation broth can be recovered by various methods, such as injecting sterile, compressed air (froth concentration); acidification; salting out; and spray drying.

Nisin Preparation is a white, free-flowing powder comprising Nisin and sodium chloride that is adjusted to an activity level of not less than 900 IU/mg by the addition of sodium chloride and nonfat milk solids.

Functional Use in Foods   Antimicrobial agent.

REQUIREMENTS

Assay Not less than 900 IU of Nisin per mg of Nisin Preparation.
Arsenic (as As) Not more than 1 mg/kg.
Differentiation of Nisin from Other Antimicrobial Substances Passes tests.
Heavy Metals (as Pb) Not more than 10 mg/kg.
Lead Not more than 2 mg/kg.
Loss on Drying Not more than 3.0%.
Microbial Limits
   Aerobic Plate Count Not more than 10 per g.
   E. coli Negative.
   Salmonella Negative.
Sodium Chloride Content Not less than 50.0%.

TESTS

Assay

Assay Medium Dissolve 10 g of bacteriological peptone, 3 g of beef extract, 3 g of sodium chloride, 1.5 g of autolyzed yeast, 1 g of brown sugar, and 15 g of agar in distilled water to a final volume of 1000 mL. Sterilize in an autoclave at 121° for 15 min. The medium can be stored in a covered container at room temperature until use. At the time of use, melt the medium and cool to approximately 50°. Add 2% of a 1:1 mixture of Tween 20 (polyoxyethylene sorbitan monolaurate) and distilled water, previously held for 20 to 30 min at 48°.

Assay Organism Maintain *Micrococcus luteus* (ATCC[3] 10240, NCIMB[4] 8166) by subculturing on agar slants of the *Assay Medium* and incubating at 30° for 48 h. Prepared slants may be stored for a maximum of 14 days at 4° until required. When required for use, the growth on the slant cultures is suspended in 7 mL of sterilized normal saline solution.

Nisin Standard Solutions Suspend 100 mg of Nisin International Reference Preparation[5] (1000 IU of Nisin per mg), accurately weighed, in 80 mL of 0.02 N hydrochloric acid. Set aside at room temperature for 2 h. Dilute the suspension to a final volume of 100.0 mL with 0.02 N hydrochloric acid. This standard stock solution contains 1000 IU of Nisin per mL. From this solution, pipet 0.5, 1.0, 2.5, 5.0, and 10.0 mL into separate 1000-mL volumetric flasks. Dilute to volume with 0.02 N hydrochloric acid to obtain *Nisin Standard Solutions* with concentrations of 0.5, 1.0, 2.5, 5.0 and 10.0 IU/mL. Store the standard stock solution for up to 7 days at 4°, or prepare a fresh solution each day.

Preparation of the Standard Curve Using normal saline solution, dilute the suspension of the *Assay Organism* to 1 in 10, and mixing thoroughly, add 2 mL of this dilution to each 100 mL of melted *Assay Medium* held at 48°. Pour the inoculated medium to a depth of 3 to 4 mm into sterile, flat-bottomed Petri dishes, and allow to solidify. Invert the plates, and store at 4° for 1 h. Bore 64 8-mm holes on 30-mm centers in each plate of the agar medium with the aid of a sterile, hollow-steel borer, 7 to 9 mm in diameter, and discard the agar discs. Transfer, in quadruplicate, 0.20-mL volumes in the concentration range of 0.5 to 10.0 IU/mL of the *Nisin Standard Solutions* into the holes. Cover the plates, and incubate them overnight at 30°. Measure the zones of inhibition to the nearest 0.1 mm by means of calipers or other appropriate devices. Plot the log of the Nisin concentration in the critical range against the zone diameters, and draw the best straight line through the plotted points.

Procedure Suspend 100 mg of the sample in 80 mL of 0.02 N hydrochloric acid in a 100-mL volumetric flask, and set aside at room temperature for 2 h. Dilute the solution to volume by adding 0.02 N hydrochloric acid. Dilute to a 1 in 200 solution with 0.02 N hydrochloric acid. Proceed as described above for the *Standard Curve*, transferring in quadruplicate a measured volume of this solution into the holes of four agar discs. After incubation, measure the zones of inhibition. From the *Standard Curve*, determine the Nisin concentrations, and average the results.

Arsenic A sample solution prepared from a 2-g sample as directed for organic compounds meets the requirements of the *Arsenic Test*, page 464, using 2 mL of the *Standard Arsenic Solution* in the control (2 μg As).

Differentiation of Nisin from Other Antimicrobial Substances

A. *Stability to Acid* Suspend a 100-mg sample in 0.02 N hydrochloric acid as described in the preparation of the *Nisin Standard Solutions* under the *Assay*. Boil this solution for 5 min. Using the *Assay* method described above, determine the Nisin concentration. No significant loss of activity is noted following this heat treatment: The calculated Nisin concentration of the boiled sample is 100 ± 5% of the *Assay* value. Adjust the pH of the Nisin solution to 11.0 by adding 5 N sodium hydroxide. Heat the solution at 65° for 30 min, and then cool. Adjust the pH to 2.0 by adding hydrochloric acid dropwise. Again determine the Nisin concentration using the *Assay* method described above. Complete loss of the antimicrobial activity of Nisin is observed following this treatment.

B. *Tolerance of* Lactococcus lactis *to High Concentrations*

---

[3]ATCC is the American Type Culture Collection, 12301 Parklawn Drive, Rockville, MD 20852.

[4]NCIMB is the National Collection of Industrial and Marine Bacteria, NCIMB Ltd., 23 St. Machar Drive, Aberdeen, Scotland AB2 1RY.

[5]The Nisin International Reference Preparation is available from the WHO International Laboratory for Biological Standards, Ministry of Agriculture, Fisheries & Food, Central Veterinary Laboratory, New Haw, Weybridge, Surrey, England.

of Nisin   Prepare cultures of *Lactococcus lactis* (ATCC 11454, NCIMB 8586) in sterile, separated milk by incubating for 18 h at 30°. Prepare one or more flasks containing 100 mL of litmus milk, and sterilize at 121° for 15 min. Suspend 0.1 g of Nisin in the sterilized litmus milk, and allow to stand at room temperature for 2 h. Add 0.1 mL of the *L. lactis* culture, and incubate at 30° for 24 h. *L. lactis* will grow in this concentration of Nisin (about 1000 IU/mL); however, it will not grow in similar concentrations of other antimicrobial substances. This test will not differentiate Nisin from subtilin.

Heavy Metals   Prepare and test a 2-g sample as directed for *Method II* under the *Heavy Metals Test*, page 513, using 20 µg of lead ion (Pb) in the control (*Solution A*).

Lead   A sample solution prepared as directed for organic compounds meets the requirements of the *Lead Limit Test*, page 518, using 2 µg of lead ion (Pb) in the control.

Loss on Drying, page 518   Dry a 2-g sample at 105° for 2 h, and continue drying to a constant weight.

Microbial Limits

    Aerobic Plate Count   Proceed as directed in Chapter 3 of the *FDA Bacteriological Analytical Manual*, Seventh Edition, 1992.

    E. coli   Proceed as directed in Chapter 4 of the *FDA Bacteriological Analytical Manual*, Seventh Edition, 1992.

    Salmonella   Proceed as directed in Chapter 5 of the *FDA Bacteriological Analytical Manual*, Seventh Edition, 1992.

Sodium Chloride Content   Prepare and test a 250-mg sample as directed for *Assay* under *Sodium Chloride*, page 282, except omit the drying. Calculate the percentage of sodium chloride in the sample.

GENERAL INFORMATION

Packaging and Storage   Store in well-closed containers at temperatures not exceeding 22°.

Paraffin, Synthetic, page 214

Change the first sentence of the *Description* to read:

    A white, practically tasteless and odorless wax that is very hard at room temperature.

Insert the following after the last sentence of the *Test* entitled *Congealing Point, Procedure*:

    Repeat the determination. If the variation is greater than 1°F, make a third determination, and record the average of the three determinations as the *Congealing Point*.

Peanut Oil (Unhydrogenated), Second Supplement, page 57

Insert the following word to precede the first word of the *Requirement* entitled *Identification*:

    Unhydrogenated

Change the *Requirement* entitled *Linolenic Acid* to read:

Linolenic Acid   Not more than 2.0%.

Replace the *Tests* entitled *Arsenic* and *Lead* to read:

Arsenic   A *Sample Solution* prepared using 4 g of sample, accurately weighed, meets the requirements of the *Arsenic Test*, page 464, using 2 mL of the *Standard Arsenic Solution* in the control (2 µg As).

Lead   Determine as directed under *Method II* in the *Atomic Absorption Spectrophotometric Graphite Furnace Method* under *Lead Limit Test*, page 168 of the Third Supplement, using a 3-g sample.

Pectin, page 215; Third Supplement, page 132

In the Third Supplement, change the *Test* entitled *Methanol, Ethanol, and Isopropanol, Standard Alcohol Solution* to read:

    Standard Alcohol Solution   Using a micropipet, transfer 50 mg each of methanol (corresponding to 39.55 µL), ethanol (corresponding to 39.47 µL), and isopropanol (corresponding to 39.28 µL) into a 1000-mL volumetric flask, dilute to volume with water, and mix.

Potassium Sorbate, page 252; First Supplement, page 9; Third Supplement, page 139

Change the *Requirement* entitled *Identification, A*, to read:

    A 1 in 10 solution responds to the flame test for *Potassium*, page 517.

Insert the following new monograph to precede the monograph entitled *Sodium Methylate*, page 291:

Sodium Metasilicate

Na$_2$O·SiO$_2$·xH$_2$O   Mol Wt (anhydrous) 122.08

INS: 550   CAS [6834-92-0]

DESCRIPTION

A hydrous (pentahydrate) or anhydrous silicate having a 1:1 molar ratio of SiO$_2$ to Na$_2$O. It occurs as a white, free-flowing granular material. At 30°, the anhydrous Sodium Metasilicate is easily soluble in water (270 g/L) as is its pentahydrate (610 g/L). The pHs of 1% solutions of anhydrous Sodium Metasilicate and of its pentahydrate are about 12.6 and 12.4, respectively.

Functional Use in Foods   Saponifying agent; boiler water additive.

REQUIREMENTS

Caution: Sodium Metasilicate and its solutions are caustic materials. Use proper protective equipment and avoid contact with the eyes, skin, and clothing. Sodium Metasilicate causes eye and skin burns. Do not inhale vapors from Sodium Metasilicate solutions.

Identification

A. Dissolve about 200 mg in 10 mL of water. Place a drop of this solution on a spot plate. Add to this one drop of 4 *M* sodium hydroxide and one drop of a solution prepared by dissolving 0.5 g of ammonium molybdate in 10 mL of water, followed by the addition of 3 mL of sulfuric acid. A deep-yellow color indicates the presence of silicate.

B. Dip a clean nichrome wire into the Sodium Metasilicate solution prepared in *Identification Test A*, and place the wire in the flame of a Bunsen burner. A bright-yellow color indicates the presence of sodium.

Assay for Silicon Dioxide and Sodium Oxide   Not less than 90.0% and not more than 110.0% of the percents claimed on the label.

Arsenic   Not more than 3 mg/kg.

Heavy Metals (as Pb)   Not more than 10 mg/kg.

Loss on Drying   Not more than 2.0% for the anhydrous and more than 42.0% for the pentahydrate.

Loss on Ignition   Not more than 0.5% for the anhydrous and between 40.5% and 42.5% for the pentahydrate.

ADDITIONAL REQUIREMENTS

Labeling   Label to indicate the percent, each, of SiO$_2$ and Na$_2$O, and whether it is anhydrous or the pentahydrate.

TESTS

Assay for Silicon Dioxide and Sodium Oxide

Silicon Dioxide   In a beaker, acidify 1 g of the sample, accurately weighed, with 5 mL of hydrochloric acid, and evaporate to dryness on a steam bath. Repeat the treatment with an additional 5 mL of hydrochloric acid, and mix the residue with 1 mL of the acid and 20 mL of water. Digest the residue on the steam bath to dissolve the soluble salts, filter the contents of the beaker through an ashless filter paper, and quantitatively transfer the residue to the paper. Wash the paper and residue thoroughly with hot water, transfer the paper to a platinum crucible, dry for 1 h at 105°, and carefully char it at low heat. Gradually increase the heat to burn away the paper, and finally ignite the crucible and its contents to constant weight at 1000°, cool in a desiccator, and weigh. Moisten the ignited residue with a few drops of water, add 15 mL of hydrofluoric acid and 5 drops of sulfuric acid (1:3). Heat the crucible with caution on a hot plate in a fume hood until all of the acid is driven off, and then ignite the residue to constant weight at a temperature of 1000°. Cool the crucible in a desiccator, and weigh. The loss in weight is equivalent to the weight of SiO$_2$ in the sample taken.

Sodium Oxide   Disperse 500 mg of the sample, accurately weighed, in 150 mL of water and heat to ensure its dissolution. Add 2 to 3 drops of phenolphthalein TS and 100.0 mL of 0.1 *N* sulfuric acid. Titrate with 0.1 *N* sodium hydroxide until a permanent pink color first appears. Subtract the volume of 0.1 *N* sodium hydroxide from the volume of 0.1 *N* sulfuric acid: Each mL of 0.1 *N* sulfuric acid is equivalent to 3.099 mg of sodium oxide.

Sample Solution for the Determination of Arsenic and Heavy Metals   Transfer 10.0 g of the sample to a 250-mL beaker, add 50 mL of 0.5 *N* hydrochloric acid, cover with a watch glass, and heat slowly to boiling. Boil gently for 15 min, cool, and let the undissolved material settle. Decant the supernatant liquid through Whatman No. 4 (or an equivalent) filter paper into a 100-mL volumetric flask, retaining as much as possible of the insoluble material in the beaker. Wash the slurry and beaker with three 10-mL portions of hot water, decanting each washing through the filter into the flask. Finally, wash the filter paper with 15 mL of hot water, cool the filtrate to room temperature, dilute to volume with water, and mix to obtain the *Sample Solution*.

Arsenic   A 10-mL portion of the *Sample Solution* meets the requirements of the *Arsenic Test*, page 464.

Heavy Metals A 20-mL portion of the *Sample Solution* meets the requirements of the *Heavy Metals Test*, page 512, using 20 µg of lead ion (Pb) in the control (*Solution A*).

Loss on Drying, page 518 Dry at 105° for 2 h. Retain the sample for the *Loss on Ignition Test*.

Loss on Ignition Ignite the sample retained from *Loss on Drying*, accurately weighed, in a suitable tared crucible at 1000° for 20 min.

GENERAL INFORMATION

Packaging and Storage Store in tight containers.

Spice Oleoresins, page 310; First Supplement, page 9; Third Supplement, page 148

Change the *Description* for various Spice Oleoresins as follows:

Oleoresin Angelica Seed Change the genus and species from *Angelica archangelica* L. to *Angelica archangelica* Linnaeus.

Oleoresin Anise Change the genus and species from *Pimpinella anisum* L. and staranise *Illicium verum* L. to anise *Pimpinella* Linnaeus and staranise *Illicium verum* Linnaeus.

Oleoresin Basil Change the genus and species from *Ocimum basilicum* L. to *Ocimum basilicum* Linnaeus.

Oleoresin Black Pepper Change the genus and species from *Piper nigrium* L. to *Piper nigrum* Linnaeus.

Oleoresin Capsicum Change the genus and species from *Capsicum frutescens* L. or *Capsicum annuum* L. to *Capsicum frutescens* Linnaeus or *Capsicum annuum* Linnaeus.

Oleoresin Caraway Change the genus and species from *Carum carvi* L. to *Carum carvi* Linnaeus.

Oleoresin Cardamom Change the genus and species from *Elettaria cardamomum* Maton to *Elettaria cardamomum* (Linnaeus) *Maton* var. minuscula Barkill.

Oleoresin Celery Change the genus and species from *Apium graveolens* L. to *Apium graveolens* Linnaeus.

Oleoresin Coriander Change the genus and species from *Coriandrum sativum* L. to *Coriandrum sativum* Linnaeus.

Oleoresin Cubeb Change the genus and species from *Piper cubeba* L. to *Piper cubeba* Linnaeus.

Oleoresin Cumin Change the genus and species from *Cuminum cyminum* L. to *Cuminum cyminum* Linnaeus.

Oleoresin Dillseed Change the genus and species from *Anethum graveolens* L. to *Anethum graveolens* Linnaeus.

Oleoresin Fennel Change the genus and species from *Foeniculum vulgare* Miller to *Foeniculum vulgare* P. Miller.

Oleoresin Ginger Change the genus and species from *Zingiber officinalis* L. Rosco to *Zingiber officinale* Roscoe.

Oleoresin Laurel Leaf Change the genus and species from *Laurus nobilis* L. to *Laurus nobilis* Linnaeus.

Oleoresin Marjoram Change the substance name to *Oleoresin Marjoram Sweet*.

Oleoresin Origanum Change the genus and species from *Origanum vulgare* L. to *Origanum vulgare* Linnaeus.

Oleoresin Paprika Change the genus and species from *Capsicum annum* L. to *Capsicum annuum* Linnaeus.

Oleoresin Parsley Leaf Change the genus and species from *Petroselinum crispum* L. to *Petroselinum crispum* (P. Miller) Nyman ex A.W. Hill.

Oleoresin Parsley Seed Change the genus and species from *Petroselinum crispum* L. to *Petroselinum crispum* (P. Miller) Nyman ex A.W. Hill.

Oleoresin Pimenta Berries Change the genus and species from *Pimenta officinalis* Lindley to *Pimenta dioica* (Linnaeus) Merrill.

Oleoresin Thyme Change the genus and species from *Thymus vulgaris* L. to *Thymus vulgaris* Linnaeus or *Thymus zygis* Linnaeus and its var. *gracelis* Boissier.

Oleoresin Turmeric Change the genus and species from *Curcuma longa* L. to *Curcuma longa* Linnaeus.

Insert the following new monograph to precede the monograph entitled *Sulfur Dioxide*, page 316:

Sucralose[6]

1,6-Dichloro-1,6-dideoxy-β-D-fructofuranosyl-4-chloro-4-deoxy-α-D-galactopyranoside; 4,1′,6′-Trichlorogalactosucrose

$C_{12}H_{19}Cl_3O_8$ Mol wt 397.64

INS: 955   CAS: [56038-13-2]

DESCRIPTION

A white to off-white, practically odorless, crystalline powder having a sweet taste. It is freely soluble in water, in methanol, and in alcohol and slightly soluble in ethyl acetate.

Functional Use in Foods   Non-nutritive sweetener; flavor enhancer.

REQUIREMENTS

Identification

A. The infrared absorption spectrum of a potassium bromide dispersion sample exhibits relative maxima (that may vary in intensity) at the same wavelengths as that of a similar preparation of Sucralose Standard for analytical use.[7]

B. The retention time of the major peak (excluding the solvent peak) in the liquid chromatogram of the *Sample Preparation* is the same as that of the *Standard Preparation* obtained in the *Assay*.

C. The R_f value of the major spot in the thin-layer chromatogram of the *Test Preparation* is the same as that of the *Standard Preparation* obtained in *Related Substances*.

Assay   Not less than 98.0% and not more than 102.0% of $C_{12}H_{19}Cl_3O_8$, calculated on the anhydrous basis.
Arsenic (as As)   Not more than 3 mg/kg.
Heavy Metals (as Pb)   Not more than 10 mg/kg.
Hydrolysis Products   Passes test.
Methanol   Not more than 0.1%.
Related Substances   Passes test.

---

[6]As of June 1, 1993, Sucralose had not been approved for use in food in the United States; however, since September 25, 1991, the use of Sucralose in food has been permitted in Canada.

[7]Available from McNeil Specialty Products Company, Regulatory Affairs Department, 501 George Street, New Brunswick, NJ 08903-2400.

Residue on Ignition   Not more than 0.7%.
Specific Rotation $[\alpha]_D^{20°}$: Between +84.0° and +87.5°, calculated on the anhydrous basis.
Water   Not more than 2.0%.

TESTS

Assay

Mobile Phase   Add 150 mL of acetonitrile (HPLC grade and filtered through a 0.45-μm filter) to 850 mL of water (glass-distilled or equivalent and filtered through a 0.45-μm filter). Mix, and de-gas thoroughly.

Chromatographic System   Fit a high-pressure liquid chromatographic system, operated at room temperature, with an 8-mm × 10-cm, 5-μm RadPakC$_{18}$ (or equivalent) reverse-phase column. Maintain the *Mobile Phase* at a pressure and flow rate (typically 1.5 mL/min) capable of giving the required elution time (see *System Suitability Test*). Use a refractive index detector.

Standard Preparation   Weigh accurately about 25 mg of Sucralose Standard for analytical use[7] into a 25-mL volumetric flask. Dissolve, and dilute to volume with *Mobile Phase*. Filter the solution through a 0.45-μm filter.

Sample Preparation   Weigh accurately about 25 mg of test sample into a 25-mL volumetric flask. Dissolve, and dilute to volume with *Mobile Phase*. Filter the solution through a 0.45-μm filter.

System Suitability Test   Chromatograph duplicate 20-μL injections of the *Standard Preparation*. Ensure that the retention time of Sucralose is approximately 9 min. It may be necessary to adjust the *Mobile Phase* composition to obtain the desired retention time. Ensure that the relative standard deviation (100 × standard deviation/mean peak area) does not exceed 2.0%.

Procedure   Analyze the *Standard Preparation* and *Sample Preparation* under the conditions described above, making duplicate 20-μL injections, and calculate the mean peak areas. Calculate the percent Sucralose from the peak areas of the *Sample Preparation* (A_U) and *Standard Preparation* (A_S) according to the following formula:

$$\% \text{ Sucralose} = 100(A_U W_S)/(A_S W_U),$$

where W_S is the weight, in mg, of the standard, and W_U is the weight, in mg, of the sample.

Arsenic   A *Sample Solution* prepared as directed for organic compounds meets the requirements of the *Arsenic Test*, page 464.

Heavy Metals   Prepare and test a 2-g sample as directed in *Method II* under the *Heavy Metals Test*, page 513, using 20 μg of lead ion (Pb) in the control (*Solution A*).

Hydrolysis Products

Spray Reagent   Dissolve 1.23 g of *p*-anisidine and 1.66 g of phthalic acid in 100 mL of methanol. Store the solution in darkness, and refrigerate to prevent it from becoming decolorized. Discard if the solution becomes discolored. (*Caution*: *p*-Anisidine is toxic if inhaled or absorbed through the skin and should be used with due caution.)

Standard Solution A Dissolve 10.0 g of mannitol, weighed to 0.001 g, in water in a 100-mL volumetric flask, and dilute to volume with water.

Standard Solution B Dissolve 40 mg of fructose and 10 g of mannitol, accurately weighed, in 25 mL of water in a 100-mL volumetric flask, and dilute to volume with water.

Sample Solution Dissolve 2.5 g of the sample in 5 mL of methanol in a 10-mL volumetric flask, and dilute to volume with methanol.

Procedure Use a thin-layer chromatographic (TLC) plate coated with a 0.25-mm layer of Merck-silica gel 60 or equivalent. Spot 5 µL of *Standard Solution A* and of *Standard Solution B* onto the plate, applying the solution slowly in 1-µL aliquots and allowing the plate to dry between applications. Spot 5 µL of the *Sample Solution* onto the plate in a similar manner. The three spots should be of similar size. Spray the plate with the *Spray Reagent*, and heat it at 100° ± 2° for 15 min. Immediately after heating, view the plate against a dark background. The spot from the *Sample Solution* is not more intense in color than the spot from *Standard Solution B* (0.1% limit).

NOTE: Darkening of the mannitol spot (the spot from *Standard Solution A*) indicates that the plate has been held too long in the oven, and a second plate should be prepared.

Methanol

Apparatus Use a suitable gas chromatographic system equipped with a hydrogen flame ionization detector and a 2.1-m × 4-mm (id) glass column packed with 80- to 100-mesh Porapak P.S. or equivalent.

Operating Conditions The operating conditions may vary, depending on the particular instrument used, but a suitable chromatogram using the above-mentioned materials may be obtained under the following conditions: column temperature: 150° (isothermal); inlet temperature: 200°; detector temperature: 250°; and carrier gas: helium, flowing at a rate of 20 mL/min. The retention time for methanol is about 2 min.

Internal Standard Solution Pipet 1.0 mL of *n*-propanol into a 100-mL volumetric flask, dilute to volume with pyridine, and mix. Transfer 5.0 mL of this solution to a 500-mL volumetric flask, dilute to volume with pyridine, and mix.

Standard Solution Pipet 2.0 mL of methanol into a 100-mL volumetric flask, dilute to volume with *Internal Standard Solution*, and mix. Transfer 1.0 mL of this solution to a 100-mL volumetric flask, dilute to volume with *Internal Standard Solution*, and mix.

Sample Solution Weigh accurately about 2 g of the sample into a 10-mL volumetric flask, dilute to volume with *Internal Standard Solution*, and mix.

Procedure Inject a 1-µL portion of the *Standard Solution* onto the gas chromatographic column, obtain the chromatogram, and measure the area of the peak produced. The relative standard deviation for replicate injections is not more than 2.0%. Calculate the mean peak areas for the *Standard Solution*. Similarly, inject a 1-µL portion of the *Sample Solution* into the gas chromatograph, and measure the areas of the peaks produced by methanol. Calculate the mean peak areas, and determine the percent of methanol in the portion of Sucralose taken using the following formula:

$$\% \text{ methanol} = (R_U/R_S)(0.158/W_S),$$

where R_U is the ratio of the peak areas of methanol to that of the internal standard obtained from the *Sample Solution*; R_S is the ratio of the peak areas of methanol to the internal standard obtained from the *Standard Solution*; the factor 0.158 is equal to the volume of methanol in the standard × dilution factor × density of methanol × 100%; and W_S is the weight, in g, of the sample.

Related Substances

Chromatographic Plates Use Whatman LKC$_{18}$ thin-layer chromatographic plates coated with a 0.20-mm layer of silica gel absorbent (or equivalent).

Mobile Phase Mix 70 volumes of 5.0% (w/v) sodium chloride in water, and add 30 volumes of acetonitrile.

Spray Reagent Use a 15% (v/v) solution of sulfuric acid in methanol.

Standard Preparations Dissolve 500.0 mg of Sucralose Standard for analytical use[6] in 5.0 mL of methanol (*Solution A*). Dilute 0.5 mL of *Solution A* with methanol to 100 mL (*Solution B*).

Test Preparation Dissolve 1.0 g of the sample in 10 mL of methanol.

Procedure Apply 5 µL each of *Solution A*, *Solution B*, and *Test Preparation* to the bottom of the chromatographic plate. Place the plate in a suitable chromatographic chamber containing freshly prepared *Mobile Phase*, and allow the solvent front to ascend approximately 15 cm. Remove the plate, allow it to dry, and spray it with the *Spray Reagent*. Heat the plate in an oven at 125° for 10 min. The main spot in the *Test Preparation* is at the same R_f value as the main spot in *Solution A*, and any other single spot in the *Test Preparation* is not more intense than the 0.5% spot in *Solution B*.

Residue on Ignition Ignite a 1- to 2-g sample as directed in *Method I*, page 533.

Specific Rotation, page 530 Determine in an aqueous solution containing 1.0 g per 100 mL, calculated on the anhydrous basis.

Water Determine by the *Karl Fischer Titrimetric Method*, page 552.

GENERAL INFORMATION

Packaging and Storage   Store in well-closed containers in a cool, dry place at less than 21°.

Sucrose, Third Supplement, page 149

Change the footnote to the NOTE located directly under *Tests* to read:

[1] International Commission for Uniform Methods of Sugar Analysis (ICUMSA), c/o British Sugar plc, Technical Centre, Colney, Norwich, England.

Triethyl Citrate, page 339; First Supplement, page 11; Second Supplement, Page 63

Change the *Description* entitled *Functional Use in Foods* to read:

Functional Use in Foods   Solvent.

Change the *Requirement* entitled *Refractive Index* to read:

Refractive Index   Between 1.439 and 1.443 at 25°; or, between 1.440 and 1.444 at 20°.

Urea, Third Supplement, page 152

Change the molecular formula to read:

CH_4N_2O

Insert the following graphic formula:

$CO(NH_2)_2$

Insert the following new monograph to precede the monograph entitled *Wintergreen Oil*, page 346:

Wheat Gluten

Vital Wheat Gluten, Devitalized Wheat Gluten

CAS: [8002-80-0]

DESCRIPTION

Wheat Gluten is the water-insoluble complex protein obtained by water extraction of wheat or wheat flour. It is soluble in alkalies and partly soluble in alcohol and dilute acids. It is a cream to light-tan, free-flowing powder. Vital Wheat Gluten is characterized by high viscoelasticity when hydrated, while devitalized Wheat Gluten has lost this character due to denaturation by heat.

Functional Use in Foods   Dough strengthener; formulation aid; nutrient supplement; processing aid; stabilizer and thickener; surface-finishing agent; and texturizing agent.

REQUIREMENTS

Identification

Add 40 mL of room-temperature water to 20 g of the sample, and stir. Vital Wheat Gluten will form a cohesive, viscoelastic mass, which can be lifted with the stirring rod without breaking apart. Devitalized Wheat Gluten will not form such a mass.

Assay   Not less than 71.0% protein, calculated on the dried basis.
Ash (Total)   Not more than 2.0%, calculated on the dried basis.
Arsenic   Not more than 1 mg/kg.
Crude Fat   Not more than 2.0%.
Heavy Metals   Not more than 5 mg/kg.
Lead   Not more than 1 mg/kg.
Loss on Drying   Not more than 10.0%.
Starch   Not more than 21.0%.

TESTS

Assay   Determine the percent of nitrogen by the *Nitrogen Determination (Kjeldahl Method)*, page 521. Percent protein equals percent $N \times 5.7$.
Ash (Total)   Proceed as directed under *Ash (Total) Method*, page 466.
Arsenic   A *Sample Solution* prepared from a 2-g sample as directed for organic compounds meets the requirements of the *Arsenic Test*, page 464, using 2 mL of the *Standard Arsenic Solution* in the control (2 µg As).

Crude Fat Proceed as directed under *Crude Fat*, in *Starches and Related Substances*, page 543.

Heavy Metals Prepare and test a 4-g sample as directed in *Method II* under the *Heavy Metals Test*, page 513, using 20 μg of lead ion (Pb) in the control (*Solution A*) and 500° as the ignition temperature.

Lead Determine as directed under *Method I* in the *Atomic Absorption Spectrophotometric Graphite Furnace Method* under the *Lead Limit Test*, page 168 of the Third Supplement, using a 1-g sample.

Loss on Drying, page 518 Dry a 2-g sample at 105° for 2 h.

Starch The remainder, after subtracting from 100.0% the sum of the percentages of *Ash (Total)*, *Loss on Drying*, and *Protein* (under *Assay*), represents the percent starch in the sample.

GENERAL INFORMATION

Packaging and Storage Store in well-closed containers.

3/ Specifications for Flavor Aromatic Chemicals and Isolates

Revised Flavor Monographs

Allyl Isothiocyanate, page 356
[FEMA No. 2034]

Change the *Assay* from M–11e to M–8a.

Delete the synonym Mustard Oil, Volatile.

Anisyl Alcohol, page 358
[FEMA No. 2099]

Change the *Refractive Index* from 1.543–1.545 to 1.542–1.547.

Benzaldehyde Glyceryl Acetal, page 66, Second Supplement
[FEMA No. 2129]

Change the *Specific Gravity* from 1.183–1.193 to 1.181–1.191.

Cinnamic Acid, page 364
[FEMA No. 2288]

Change the synonym from 3-Phenylpropionic Acid to 3-Phenylpropenoic Acid.

1-Decanol, Natural, page 368; page 14, First Supplement (Decyl Alcohol, Alcohol C-10)
[FEMA No. 2365]

Change the name to *Decyl Alcohol*.

Change the synonym from Decyl Alcohol to 1-Decanol.

Keep the synonym Alcohol C-10.

Delete Natural.

Ethyl Methylphenylglycidate, page 376
[FEMA No. 2444]

Change the *Specific Gravity* from 1.086–1.112 to 1.086–1.096.

Change the *Refractive Index* from 1.504–1.513 to 1.503–1.509.

Change the *Acid Value* from 2.0 max. to 1.0 max.

Ethyl 3-Methylthiopropionate, page 160, Third Supplement
[FEMA No. 3343]

Insert the following under *Physical Form/Odor*: Colorless to pale yel liq/onion-like, fruity, sweet.

Linalool, page 392
[FEMA No. 2635]

Under *Other Requirements*, page 393, delete *Angular Rotation*.

1-Octanol, Natural, page 406; page 19, First Supplement (Alcohol C-8, Octyl Alcohol, Capryl Alcohol)
[FEMA No. 2800]

Change the name to *Octyl Alcohol*.

Change the synonym from Octyl Alcohol to 1-Octanol.

Keep the synonyms Alcohol C-8 and Capryl Alcohol.

Delete Natural.

Terpinyl Propionate, page 416
[FEMA No. 3053]

Change the *Specific Gravity* from 0.944–0.949 to 0.947–0.952.

Change the *Refractive Index* from 1.461–1.466 to 1.462–1.468.

New Flavor Monographs

General Information and Description

Name of Substance (Synonyms)	Mol Wt/Formula/ Structure	Physical Form/Odor[1]	Solubility/ B.P.	GLC Profile	Solubility in Alcohol
Allyl Isovalerate [FEMA No. 2045]	142.20/$C_8H_{14}O_2$/ $(CH_3)_2CHCH_2CO_2CH_2CH=CH_2$	colorless to pale yel liq/ fruit-like, apple aroma			
Ethyl 10-Undecenoate [FEMA No. 2461]	212.33/$C_{13}H_{24}O_2$/ $H_2C=CH(CH_2)_8CO_2C_2H_5$	colorless to pale yel liq	258°–259°		
2-Mercaptopropionic Acid [FEMA No. 3180]	106.16/$C_3H_6O_2S$/ $CH_3CH(SH)COOH$	colorless to pale yel liq/ roasted, meaty odor	m-water, alc, ether, acetone/117°		
Methyl-3-methylthiopropionate [FEMA No. 2720]	134.19/$C_5H_{10}O_2S$/ $CH_3SCH_2CH_2CO_2CH_3$	colorless to pale yel liq/ onion-like	74°–75°		
Myristyl Alcohol (1-Tetradecanol, Tetradecyl Alcohol)	214.38/$C_{14}H_{30}O$/ $CH_3(CH_2)_{12}CH_2OH$	colorless to white waxy solid flakes/ waxy	s-ether, ss-alc, ins-water/167°		
δ-Undecalactone (5-Hydroxyundecanoic Acid Lactone) [FEMA No. 3294]	184.28/$C_{11}H_{20}O_2$	colorless to pale yel liq/ creamy, peach-like	152°–155°		

NOTES:

[1] cryst = crystal or crystalline; liq = liquid; NLT = not less than; yel = yellow.

[2] Sample weight and equivalence factor are given for methods M-2a, M-4, and M-6; sample weights (g) and amounts (mg) of substance equivalent to 0.5 N sodium hydroxide are given for method M-11a.

Requirements

I.D. Test	Assay Min, %[2]	A.V. Max	Ref. Index	Sp. Gr.	Other Requirements
	98.0% min (M–6)	1.0	1.413–1.418	0.879–0.884	
	98.0% min (M–6)	1.0	1.436–1.440	0.877–0.879	
	98.0% min (M–11a; 1.0 g/53.08)		1.479–1.484	1.192–1.200	
	99.0% min (M–8a)	1.0	1.462–1.468	1.069–1.078	
	98.0% as $C_{14}H_{30}O$ (M–12; 1.4 g/7.2)	1.0			**Melting Range:** 38°–41°; **Iodine Value:** 3.0 max; **Saponification Value:** 1.0 max
	98.0% min (M–8a)		1.457–1.461	0.956–0.961	

4/ Test Methods for Flavor Aromatic Chemicals and Isolates

No change.

5/ GLC Analysis of Flavor Aromatic Chemicals and Isolates

No change.

6/ General Tests and Apparatus

Oil Content of Synthetic Paraffin, page 525

Insert the following into the *Procedure*, last paragraph, as the penultimate sentence:

Determine the weight of the oil residue, in g, by subtracting the weight of the empty stoppered bottle from the weight of the stoppered bottle plus the oil residue after the evaporation procedure, and record the results as A (see *Calculation*).

7/ Solutions and Indicators

No change.

8/ General Information

No change.

9/ Infrared Spectra

Morpholine

Insert the infrared spectrum of *Morpholine* to precede the spectrum for *Paraffin, Synthetic*, page 718.

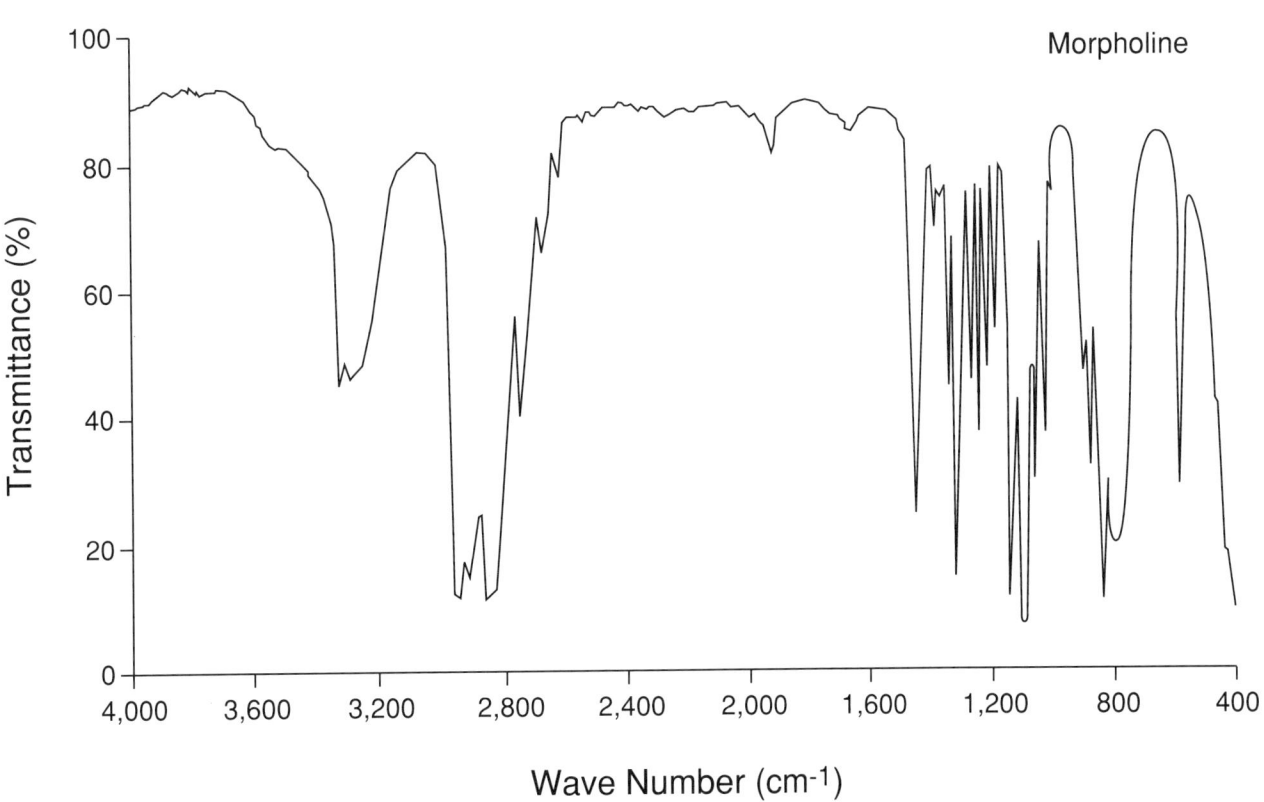

Index

Page citations refer to the First (1–34), Second (35–90), Third (91–185), and Fourth (186–226) Supplements to the Third Edition of the *Food Chemicals Codex*. An asterisk (*) indicates a new listing.

Aceteugenol, 15
3-Acetyl-2,5-dimethyl Furan, 12, 64
Acetyl Eugenol, 15
N-Acetyl-L-Methionine, 97, 189
2-Acetyl Pyrazine, 66, 178
Achilleic Acid, 97
Acid Hydrolyzed Proteins, 3
Acid-Insoluble Matter in Anhydrous Calcium Chloride, 70
Aconitic Acid, 97
DL-Alanine, 98
L-Alanine, 99
Alcohol C-6, 17
Alcohol C-8, 19, 213
Alcohol C-10, 14, 212
Alcohol C-12, 17
Aldehyde C-7, 16
Aldehyde C-14, 157
Allura Red AC, 48
p-Allylanisole, 15, 65
Allyl Caproate, 12
4-Allylguaiacol, 15
Allyl Heptanoate, 22
Allyl Heptoate, 22
Allyl Hexanoate, 12
Allyl Ionone, 12
Allyl α-Ionone, 12
Allyl Isothiocyanate, 212
*Allyl Isovalerate, 214
4-Allyl-2-methoxyphenol, 15

4-Allyl-2-methoxyphenyl Acetate, 15
Aluminum Ammonium Sulfate, 4
Aluminum Potassium Sulfate, 4
Aluminum Sodium Sulfate, 5, 99
Aluminum Sulfate, 5
*Ammoniated Glycyrrhizin, 189
Ammonium Bicarbonate, 5
Ammonium Glycyrrhizinate, 200
Ammonium Glycyrrhizinate, Pentahydrate 200
Amyl Butyrate, 17
Amylcinnamaldehyde, 12
α-Amylcinnamaldehyde, 12
Amyl Heptanoate, 66
Amyl Salicylate, 17
Amyl Vinyl Carbinol, 162
Anethole, 64
Angelica Seed, Oleoresin, 9, 207
Anhydrous Calcium Chloride, 38
p-Anisaldehyde, 18
Anise, Oleoresin, 10, 207
Anisic Aldehyde, 18
Anisyl Acetate, 12
Anisylacetone, 66
Anisyl Alcohol, 212
Anisyl Formate, 158
Annatto Extracts, 37, 99, 190
L-Arginine, 99, 190
L-Arginine Monohydrochloride, 99
L-Asparagine, 99

Aspartame, 5, 31
DL-Aspartic Acid, 100
L-Aspartic Acid, 100
Assay by Determination of Aldehydes and Ketones—Hydroxylamine Method, 24
Assay by Determination of Esters, 24
Assay by Determination of Esters, High-Boiling Method, 24
Assay by Gas–Liquid Chromatography, 25
Assay for Determination of Phenols, 25
Atomic Absorption Spectrophotometry (Graphite Furnace Method), 168
Autolyzed Yeast Extract, 5
BHA, 38
Barium Hydroxide, 0.2 N, 88
Basil, Oleoresin, 10, 207
Bay Oil, 6
Beet Sugar, 149
Benzaldehyde, 13
Benzaldehyde Glyceryl Acetal, 66, 212
Benzene (in Paraffinic Hydrocarbon Solvents), 71
Benzodihydropyrone, 22
Benzyl Acetate, 13
Benzyl Butyrate, 26
Benzyl Cinnamate, 13, 156

Benzyl Formate, 158
Benzyl Isobutyrate, 156
Benzyl Isovalerate, 156
Benzyl Propionate, 26
Benzyl Salicylate, 13
Black Pepper Oil, 100
Black Pepper, Oleoresin, 207
Brilliant Blue FCF, 43
Brominated Vegetable Oil, 38
Butane, 100
n-Butane, 100
1-Butanol, 64
Butter Starter Distillate, 149
Butyl Acetate, 13
n-Butyl Acetate, 13
Butyl Alcohol, 64
Butyl Aldehyde, 13
Butylated Hydroxyanisole, 38
Butyl Isovalerate, 22
Butyl Octadecanoate, 158
Butyl Phenylacetate, 66
Butyl Stearate, 158
Butyraldehyde, 13
Calcium Carbonate, 38.
Calcium Chloride, 102
Calcium Chloride, Anhydrous, 38
Acid-Insoluble Matter in, 70
Calcium Gluconate, 102
Calcium Oxide, 6
Calcium Pantothenate, 103
Calcium Pantothenate, Calcium Chloride Double Salt, 103, 190
Calcium Pantothenate, Racemic, 103, 191
Calcium Sorbate, 104
Calcium Sulfate, 39
Calcium Sulfate TS, 29
Camphene, 13
D-Camphor, 158
Cane Sugar, 149
Canola Oil, 104
Caproic Aldehyde, 16
Capryl Alcohol, 19, 213
Capsicum, Oleoresin, 207
Caraway, Oleoresin, 10, 207
Carbamide, 152
Carbon, Activated, 6, 106
Cardamom, Oleoresin, 10, 207
Carmine, 6
Carnauba Wax, 107
Carrageenan, 39
Carrot Seed Oil, 107
Carvacrol, 13
L-Carveol, 158
L-Carvyl Acetate, 158

β-Caryophyllene, 13
Casein and Caseinate Salts, 39
Celery, Oleoresin, 207
Celery Seed Oil, 6
Chromatography, 27, 73
Cinnamal, 13
Cinnamaldehyde, 13
Cinnamic Acid, 212
Cinnamic Aldehyde, 13
Cinnamyl Acetate, 14
Cinnamyl Butyrate, 158
Cinnamyl Cinnamate, 158
Cinnamyl Formate, 14, 64
Cinnamyl Isobutyrate, 158
Cinnamyl Propionate, 64
Citric Acid, 107
Citridic Acid, 97
Citronellal, 64
Citronellol, 14
Citronellyl Acetate, 14
Citronellyl Formate, 14
*Cocoa Butter Substitute, 191
Coconut Oil (Unhydrogenated), 40
Colors, FD&C (also see specific color under FD&C), 73
   Chromium, 73
   Ether Extracts, 74
   Leuco Base, 74
   Mercury, 75
   Sodium Chloride, 76
   Sodium Sulfate, 76
   Total Color, 76
   Uncombined Intermediates and Products of Side Reactions, 77
   Volatile Matter, 80
   Water-Insoluble Matter, 80
Copper Sulfate, 108
Coriander, Oleoresin, 10, 207
Corn Oil (Unhydrogenated), 41
Corn Syrup, 112
Corn Syrup, High-Fructose, 51
Cottonseed Oil (Unhydrogenated), 42, 193
Cubeb, Oleoresin, 10, 207
Cumaldehyde, 14, 64
Cumin, Oleoresin, 10, 207
Cuminal, 14, 64
Cuminic Aldehyde, 14, 64
p-Cuminic Aldehyde, 14, 64
Cupric Citrate TS, Alkaline, 173
Cupric Sulfate, 108
Cyclamen Aldehyde, 14, 65
p-Cymene, 66, 179
L-Cysteine Monohydrochloride, 109
L-Cystine, 109

δ-Decalactone, 14, 65
γ-Decalactone, 158
1-Decanol, Natural, 14, 212
cis-4-Decen-1-al, 65
$trans$-2-Decen-1-al, 65
Decyl Alcohol, 14, 212
Detector Tubes, 173
Devitalized Wheat Gluten, 210
Dextrose, 109
Diatomaceous Earth, 110
Dibenzyl Ether, 22
1,6-Dichloro-1,6-dideoxy-β-D-fructofuranosyl-4-chloro-4-deoxy-α-D-galactopyranoside, 208
Diethylene Imidoxide, 201
Diethylene Oximide, 201
Diethyl Succinate, 15
Dillseed, Oleoresin, 10, 207
1,2-Dimethoxy-4-allylbenzene, 19
2,5-Dimethyl-3-acetylfuran, 12, 64
Dimethyl Benzyl Carbinyl Butyrate, 15
2,6-Dimethyl-5-heptenal, 66, 156
3,7-Dimethyl-2,6-octadien-1-yl Acetate, 16
cis-3,7-Dimethyl-2,6-octadien-1-yl Acetate, 162
3,7-Dimethyl-2,6-octadien-1-yl Benzoate, 16
3,7-Dimethyl-2,6-octadien-1-yl Butyrate, 16
3,7-Dimethyl-2,6-octadien-1-yl Formate, 16
3,7-Dimethyl-2,6-octadien-1-yl Phenylacetate, 16
3,7-Dimethyl-2,6-octadien-3-yl Propionate, 18
3,7-Dimethyl-3-octanol, 20
3,7-Dimethyl-6-octen-1-al, 14
3,7-Dimethyl-6-octen-1-ol, 14
3,7-Dimethyl-6-octen-1-yl Acetate, 14
3,7-Dimethyl-6-octen-1-yl Formate, 14
α,α-Dimethylphenethyl Butyrate, 15
Dimethylpolysiloxane, 43
Dimethyl Sulfide, 158
Disodium Guanylate, 6
δ-Dodecalactone, 15, 65
γ-Dodecalactone, 158
1-Dodecanol, 17
Dried Glucose Syrup, 112
Enocianina, 7
Enzyme-Modified Fat, 193

*Enzyme-Modified Milkfat, 193
Enzyme Preparations, 6
Epsom Salt, 8
Equisetic Acid, 97
Erythrosine, 46
Estragole, 15, 65, 156
Ethoxyquin, 110
Ethyl Butyl Ketone, 16
Ethyl Cinnamate, 15
Ethylene Brassylate, 160
2-Ethyl Fenchol, 65
Ethyl Formate, 156
4-Ethylguaiacol, 160
Ethyl Isobutyrate, 22
Ethyl 2-Methylbutyrate, 15
Ethyl Methylphenylglycidate, 212
Ethyl-3-Methylthiopropionate, 160, 213
Ethyl Myristate, 22
Ethyl 9-Octadecenoate, 160
Ethyl Oleate, 160
Ethyl 3-Phenylpropenate, 15
*Ethyl 10-Undecenoate
Eugenic Acid, 15
Eugenol, 15
Eugenol Acetate, 15
Eugenyl Acetate, 15
Eugenyl Methyl Ether, 19
Extractable Organic Compounds (in Hydrochloric Acid), 80
FCC Operating Procedures, 175
    Criteria for *Food Chemicals Codex* Grade, 176
    Functions of the Committee on Food Chemicals Codex, 175
    Further Information, 177
    Organization, 175
    Procedure for Revising Specifications, 177
    Procedures for Submission and Development of Specifications, 176
    Requirements for Listing Substances in the *Food Chemicals Codex*, 178
FD&C Blue No. 1, 43
FD&C Blue No. 2, 44
FD&C Green No. 3, 45
FD&C Red No. 3, 46
FD&C Red No. 40, 48
FD&C Yellow No. 5, 49
FD&C Yellow No. 6, 50
Farnesol, 15
Fast Green FCF, 45
Fats and Related Substances, 82
    Cold Test, 82
    Fatty Acid Composition, 82
    Melting Range I, 82
    Stability (Active Oxygen Method), 83
Fennel Oil, 110
Fennel, Oleoresin, 10, 207
Ferrous Fumarate, 110
Fructose, 7, 110
Fuchsin–Sulfurous Acid TS, 173
Fully Hydrogenated Rapeseed Oil, 140
Fusel Oil Refined, 66
4-O-β-Galactopyranosyl-D-glucose, 194
Garlic Oil, 111
Gas Chromatographic Analysis of Butyl and Isobutyl Alcohols, 68
Gas Chromatography, 27
Gellan Gum, 111
Geranyl Acetate, 16
Geranyl Benzoate, 16
Geranyl Butyrate, 16
Geranyl Formate, 16
Geranyl Phenylacetate, 16
Ginger, Oleoresin, 207
Glucose Syrup, 112
Glucose Syrup, Dried, 112
Glucose Syrup Solids, 112
L-Glutamic Acid, 113, 194
L-Glutamic Acid Hydrochloride, 113
L-Glutamine, 114
Glutaral, 114
Glutaraldehyde, 114
Glyceryl Behenate, 115
Glyceryl-Lacto Esters of Fatty Acids, 122
Glyceryl Monostearate, 116, 194
Glyceryl Tribehenate, 115
Glyceryl Tridocosanoate, 115
Glyceryl Tripropanoate, 160
Glycine, 118
Granulated Sugar, 149
Grape Skin Extract, 7
Gum Ghatti, 118
Helium, 119
Heptaldehyde, 16
Heptanal, 16
2-Heptanone, 16
3-Heptanone, 16
2,4-Hexadienoic Acid, Calcium Salt, 104
Hexahydropyridine, 162
γ-Hexalactone, 160
Hexaldehyde, 16
Hexanal, 16
Hexanes, 50
1-Hexanol, 17
cis-3-Hexen-1-ol, 16
cis-3-Hexenyl Acetate, 160
trans-2-Hexenyl Acetate, 160
n-Hexyl Acetate, 22
Hexyl Alcohol, Natural, 17
Hexyl Isovalerate, 17
High-Fructose Corn Syrup, 51, 119
    Solids in, 84
L-Histidine, 120
L-Histidine Monohydrochloride, 120
Hydrocarbons, Mixed Paraffinic, 50
Hydrochloric Acid, 52
    Extractable Organic Compounds in, 80
Hydrolyzed Milk Protein, 3
Hydrolyzed Plant Protein (HPP), 3
Hydrolyzed Vegetable Protein (HVP), 3
4-Hydroxydecanoic Acid Lactone, 158
4-Hydroxy-2,5-dimethyl-3(2H)furanone, 160
4-Hydroxydodecanoic Acid Lactone, 158
4-Hydroxyhexanoic Acid Lactone, 160
4-Hydroxy-3-methoxy-ethylbenzene, 160
5-Hydroxynonanoic Acid, Lactone, 66
5-Hydroxyoctanoic Acid, Lactone, 66
4-(p-Hydroxyphenyl)-2-butanone, 22
2-Hydroxy-Propanoic Acid Monopotassium Salt, 137
2-Hydroxy-Propanoic Acid Monosodium Salt, 144
5-Hydroxyundecanoic Acid Lactone, 214
Indian Gum, 118
Indigo Carmine, 44
Indigotine, 44
Indigotine Disulfonate, 44
Invert Sugar, 53, 84, 120, 172, 194
Invert Sugar Syrup, 53
Iodine Value, 169
α-Ionone, 17
β-Ionone, 17
Isoamyl Benzoate, 66
Isoamyl Butyrate, 17, 156
Isoamyl Formate, 26, 156
Isoamyl Salicylate, 17

Isoborneol, 160
Isobornyl Acetate, 65
Isobutane, 121
Isobutyl Acetate, 17
Isobutyl Alcohol, 17, 65
Isobutyric Acid, 17
DL-Isoleucine, 121
L-Isoleucine, 7, 121
Isophenylformic Acid, 17
p-Isopropylbenzaldehyde, 14, 64
Isovaleric Acid, 156
Konjac, 122
Konjac Flour, 122
Konjac Gum, 122
Konnyaku, 122
Lactated Mono-Diglycerides, 122
Lactic Acid, 123
*Lactose, 85, 194
Lard (Unhydrogenated), 54
Laurel Leaf, Oleoresin, 10, 207
Lauryl Alcohol, Natural, 17
LEAR, 104
Lemon Oil, Coldpressed, 195
DL-Leucine, 123
L-Leucine, 7, 124
Limit Test for Phenolic Impurities, 25
Linalool, 213
Linalyl Propionate, 18
*Linoleic Acid, 195
Locust (Carob) Bean Gum, 196
Low Erucic Acid Rapeseed Oil, 104
L-Lysine Monohydrochloride, 124, 196
Magnesium Chloride, 124
Magnesium Oxide, 7
Magnesium Sulfate, 8
Maltodextrin, 125
Malt Extract, 196
*Malt Syrup, 196
Mandarin Oil, Coldpressed, 8
Marjoram, Oleoresin, 10, 207
p-Mentha-6,8-dien-2-ol, 158
p-Mentha-6,8-dien-2-yl Acetate, 158
3-p-Menthanol, 18
l-p-Menthan-3-one, 18
dl-p-Menthan-3-yl Acetate, 18
l-p-Menthan-3-yl Acetate, 18
Menthen-1-ol-8, 20, 65
Menthol, 18
l-Menthone, 18
dl-Menthyl Acetate, 18
l-Menthyl Acetate, 18
*2-Mercaptopropionic Acid, 214
Methional, 162

DL-Methionine, 125
L-Methionine, 125
p-Methoxybenzaldehyde, 18
p-Methoxybenzyl Acetate, 12
p-Methoxybenzyl Formate, 158
4-p-Methoxyphenyl-2-butanone, 66
2-Methoxypyrazine, 18
4'-Methyl Acetophenone, 18
Methyl Amyl Ketone, 16
p-Methylbenzaldehyde, 164
Methyl Benzoate, 90
Methylbenzyl Acetate, 18
2-Methylbutyl Isovalerate, 19, 65
2-Methylbutyl-3-methylbutanoate, 19, 65
α-Methylcinnamaldehyde, 19
6-Methylcoumarin, 66, 179
Methyl Eugenol, 19
Methyl Formate, 55
Methyl Heptyl Ketone, 162
Methyl Hexyl Ketone, 66
2-Methyl-3-(p-isopropylphenyl)-propionaldehyde, 14, 65
Methyl 2-Methylbutanoate, 19
Methyl 2-Methylbutyrate, 19
*Methyl-3-methylthiopropionate, 214
Methyl Nonyl Ketone, 164
2-Methylpentanoic Acid, 160
4-Methylpentanoic Acid, 160
2-Methyl-2-pentenoic Acid, 160
2-Methyl Propanoic Acid, 17
Methyl Salicylate, 19
Methyl Sulfide Thiobismethane, 158
3-Methylthiopropionaldehyde, 162
Methyl p-Tolyl Ketone, 18
Mixed Paraffinic Hydrocarbons, 50
Monoammonium L-Glutamate, 126
*Monoammonium Glycyrrhizinate, 200
Monopotassium L-Glutamate, 126
Monosodium L-Glutamate, 126, 201
Monostearin, 116
*Morpholine, 201, 221
*Mustard Oil, 202
Myristaldehyde, 162
*Myristyl Alcohol, 214
Natamycin, 126
Neryl Acetate, 162
*Nickel, 202
Nickel Catalysts, 202
*Nisin Preparation, 203
Nitrogen, 128
Nitrogen Enriched Air, 128
Nitrogen Oxide, 129
Nitrous Oxide, 129

δ-Nonalactone, 66
2-Nonanone, 162
(Z,Z)-9,12-Octadecadienoic Acid, 195
δ-Octalactone, 66
1-Octanol, Natural, 19, 213
2-Octanone, 66
1-Octene-3-ol, 162
1-Octen-3-yl Acetate, 19
1-Octen-3-yl Butyrate, 19
3-Octyl Acetate, 19
Octyl Alcohol, 19, 213
Octyl Formate, 19
Octyl Isobutyrate, 162
Octyl 2-Methylpropanoate, 162
Oil Content of Synthetic Paraffin, 218
Oleoresin Angelica Seed, 207
Oleoresin Anise, 207
Oleoresin Basil, 207
Oleoresin Black Pepper, 207
Oleoresin Capsicum, 207
Oleoresin Caraway, 207
Oleoresin Cardamom, 207
Oleoresin Celery, 207
Oleoresin Coriander, 000
Oleoresin Cubeb, 207
Oleoresin Cumin, 207
Oleoresin Dillseed, 207
Oleoresin Fennel, 207
Oleoresin Ginger, 207
Oleoresin Laurel Leaf, 207
Oleoresin Marjoram, 207
Oleoresin Origanum, 207
Oleoresin Paprika, 207
Oleoresin Parsley Leaf, 207
Oleoresin Parsley Seed, 207
Oleoresin Pimenta Berries, 207
Oleoresins, Spice, 9, 10, 207
Oleoresin Thyme, 207
Oleoresin Turmeric, 207
Operating Procedures, FCC, 175
   Criteria for *Food Chemicals Codex* Grade, 176
   Functions of the Committee on Food Chemicals Codex, 175
   Further Information, 177
   Organization, 175
   Procedure for Revising Specifications, 177
   Procedures for Submission and Development of Specifications, 176
   Requirements for Listing Substances in the *Food Chemicals Codex*, 176

Origanum, Oleoresin, 10, 207
Ox Bile Extract, 130
Ozone, 131
Palmarosa Oil, 132
Palm Kernel Oil (Unhydrogenated), 55
Palm Oil (Unhydrogenated), 56
Paprika, Oleoresin, 207
Paraffinic Hydrocarbons, Mixed, 50
Paraffin, Synthetic, 205
Parsley Leaf, Oleoresin, 10, 207
Parsley Seed, Oleoresin, 10, 207
Peanut Oil (Unhydrogenated), 57, 205
Pectin, 132, 205
1,5-Pentanedial, 114
Peppermint Oil, 8
Perlite, 135
Petroleum Wax, Synthetic, 8
Phenethyl Isovalerate, 19
2-Phenethyl 2-Methylbutyrate, 20
Phenoxyethyl Isobutyrate, 20
Phenylacetaldehyde, 20
DL-Phenylalanine, 135
L-Phenylalanine, 135
Phenylethyl Anthranilate, 162
Phenylethyl Butyrate, 162
Pimaricin, 126
Pimenta Berries, Oleoresin, 10, 207
α-Pinene, 157
β-Pinene, 157
Piperidine, 162
Poloxamer 331, 57
Poloxamer 407, 57
Polydextrose, 57, 136
Polydextrose Solution, 59
Potassium Alginate, 8
Potassium Benzoate, 136
Potassium Bicarbonate, 8, 137
Potassium Carbonate, 8
Potassium Chloride, 137
Potassium Lactate Solution, 137
Potassium Nitrate, 9
Potassium Sorbate, 9, 139, 205
L-Proline, 139
Propane, 140
1,2,3-Propanetriol Octadecanoate, 116
p-Propenylanisole, 64
Purified Oxgall, 130

Rapeseed Oil, Fully Hydrogenated, 140
Rapeseed Oil, Superglycerinated, 141
Reducing Sugars Assay, 169
Rhodinyl Acetate, 20
Safflower Oil (Unhydrogenated), 60
DL-Serine, 142
L-Serine, 9, 142
Silicon Dioxide, 9
Sodium Alginate, 9
Sodium Aluminosilicate, 143
Sodium Bicarbonate, 9
Sodium Carbonate, 9
Sodium Choleate, 130
Sodium Lactate Solution, 144
Sodium Magnesium Aluminosilicate, 145
*Sodium Metasilicate, 206
Sodium Saccharin, 9
Sodium Stearyl Fumarate, 147
Sodium Thiosulfate, 88
Sorbitol, 147
Sorbitol Solution, 148
Soybean Oil (Unhydrogenated), 61
Spice Oleoresins, 9, 148, 207
Spike Lavender Oil, 10
Starter Distillate, 149
*Sucralose, 208
Sucrose, 149, 210
Sugar, 149
Sulfur (by Oxidative Microcoulometry), 85
Sulfur Dioxide Determination, 170
Sulfuric Acid, 151
Sunflower Oil, 62
Sunset Yellow FCF, 50
Superglycerinated Fully Hydrogenated Rapeseed Oil, 141
TBHQ, 63
Tallow, 63
Tartrazine, 49
Terpineol, 20, 65
Terpinyl Propionate, 213
Test for Free Phenols, 25
Test for Phenols Using Cassia Flask Method, 25
Test Solutions (TS) and Other Reagents, 29
Tetradecanal, 162

1-Tetradecanol, 214
Tetradecyl Alcohol, 214
Tetrahydrofurfuryl Alcohol, 162
Tetrahydrolinalool, 20
Tetrahydro-2H-1, 4-oxazine, 201
Thiamine Mononitrate, 32
L-Threonine, 151
Thyme, Oleoresin, 10, 207
Thymol, 162
d-α-Tocopheryl Acetate Concentrate, 10
Tolualdehyde (mixed isomers), 162
para-Tolualdehyde, 164
α-Toluic Aldehyde, 20
Tolyl Acetate (So Called), 18
p-Tolyl Aldehyde, 164
Tolyl Aldehyde (mixed isomers, Methyl Benzaldehyde), 162
Triacetin, 11
Triatomic Oxygen, 131
4,1′,6′-Trichlorogalactosucrose, 208
Triethyl Citrate, 11, 63, 210
Trimethylamine, 164
4(2,6,6-Trimethyl-1-cyclohexenyl)-3-butene-2-one, 17
4(2,6,6-Trimethyl-2-cyclohexenyl)-3-butene-2-one, 17
3,7,11-Trimethyl-2,6,10-dodecatrien-1-ol, 15
Tripropionin, 160
DL-Tryptophan, 151
L-Tryptophan, 152
Turmeric, Oleoresin, 207
L-Tyrosine, 152
*δ-Undecalactone, 214
γ-Undecalactone, 157
2-Undecanone, 164
Urea, 152, 210
Valeraldehyde, 164
γ-Valerolactone, 20
L-Valine, 153
Vital Wheat Gluten, 210
*Wheat Gluten, 210
Xanthan Gum, 153
Xylitol, 11, 153
Zein, 154
Zinc Gluconate, 155
Zinc Sulfate, 11